THE JUSTICE INNOVATION CENTER

Identifying the Needs and Challenges of Criminal Justice Agencies in Small, Rural, Tribal, and Border Areas

Jessica Saunders, Meagan Cahill,
Andrew R. Morral, Kristin J. Leuschner,
Gregory Midgette, John S. Hollywood,
Mauri Matsuda, Lisa Wagner,
Jirka Taylor

For more information on this publication, visit www.rand.org/t/rr1479

Library of Congress Cataloging-in-Publication Data is available for this publication.
ISBN: 978-0-8330-9657-9

Published by the RAND Corporation, Santa Monica, Calif.
© Copyright 2016 RAND Corporation
RAND® is a registered trademark.

Cover: (Background photo) City of Woodstock, Illinois; (fingerprint) Andrey Prokhorov/GettyImages

Limited Print and Electronic Distribution Rights

This document and trademark(s) contained herein are protected by law. This representation of RAND intellectual property is provided for noncommercial use only. Unauthorized posting of this publication online is prohibited. Permission is given to duplicate this document for personal use only, as long as it is unaltered and complete. Permission is required from RAND to reproduce, or reuse in another form, any of its research documents for commercial use. For information on reprint and linking permissions, please visit www.rand.org/pubs/permissions.

The RAND Corporation is a research organization that develops solutions to public policy challenges to help make communities throughout the world safer and more secure, healthier and more prosperous. RAND is nonprofit, nonpartisan, and committed to the public interest.

RAND's publications do not necessarily reflect the opinions of its research clients and sponsors.

Support RAND
Make a tax-deductible charitable contribution at
www.rand.org/giving/contribute

www.rand.org

This report describes the RAND Corporation's activities under award 2014-MU-CX-K003, the National Law Enforcement and Corrections Technology Center (NLECTC) System: Small, Rural, Tribal, and Border (SRTB) Regional Center. In year 1, the center was renamed the Justice Innovation Center for SRTB Criminal Justice Agencies. The opinions, findings, and conclusions or recommendations expressed in this publication are those of the authors and do not necessarily reflect those of the Department of Justice.

This work is dedicated to the memory of our friend and colleague
Marie Griffin
1967–2016

Preface

This report describes the RAND Corporation's activities under the Justice Innovation Center for Small, Rural, Tribal, and Border Criminal Justice Agencies.

This report should be of interest to policymakers and practitioners interested in the unmet operational needs of small, rural, tribal, and border criminal justice agencies, as well as to the larger criminal justice community and associated research community.

RAND Justice Policy

The research reported here was conducted in the RAND Justice Policy Program, which spans both criminal and civil justice systems issues with such topics as public safety, effective policing, police–community relations, drug policy and enforcement, corrections policy, use of technology in law enforcement, tort reform, catastrophe and mass-injury compensation, court resourcing, and insurance regulation. Program research is supported by government agencies, foundations, and the private sector.

This program is part of RAND Justice, Infrastructure, and Environment, a division of the RAND Corporation dedicated to improving policy- and decisionmaking in a wide range of policy domains, including civil and criminal justice, infrastructure protection and homeland security, transportation and energy policy, and environmental and natural resource policy.

Questions or comments about this report should be sent to Meagan Cahill (Meagan_Cahill@rand.org). For more information about RAND Justice Policy, see www.rand.org/jie/justice-policy or contact the director at justice@rand.org.

Contents

Preface v
Figures and Tables ix
Summary xi
Acknowledgments xxv
Abbreviations xxvii

CHAPTER ONE
Introduction 1
How Are SRTB Agencies Different? 1
The Justice Innovation Center's Four Research Goals 9
Organization of This Report 12

CHAPTER TWO
Methodology 15
Defining Communities of Practice: SRTB Criminal Justice Agencies 15
Literature Review Approach 19
Semistructured Interviews 20
Advisory Panel 27

CHAPTER THREE
Literature Review 29
General Challenges Facing SRTB Agencies 29
Law Enforcement Technology Needs Among SRTB Agencies 31
Court Technology Needs Among SRTB Agencies 36
Corrections Technology Needs Among SRTB Agencies 37
Conclusion 38

CHAPTER FOUR
Agency Interviews 41
Geographic Challenges 42
Funding Challenges 44
Social Service Provision Challenges 47
Personnel Challenges 50
Communications and Information-Technology Management 54
Data-Sharing and Interoperability 58
Crime-Specific Challenges 64

Infrastructure Challenges 65
Resistance to Technological Change.. 67
Legal and Policy Barriers... 69
Conclusion... 70

CHAPTER FIVE
The Justice Innovation Center Advisory Panel: Identifying Science- and Technology-Related Needs to Address Pressing Issues... 73
Needs That the Advisory Panel Identified... 74
Key Themes from the Panel Discussions... 75
Conclusion... 83

CHAPTER SIX
Conclusions... 85
Understanding Technology in SRTB Contexts 85
Gathering Input on Operational Needs from the Field: Interviews 86
Gathering Input on Operational Needs from the Field: Advisory Panel...................... 87
Synthesis of Lessons from the Justice Innovation Center's Initial Work 88
Plans for Future Years 88

APPENDIXES
A. Letter of Invitation to the Study, Letter of Support for the Study, and Justice Innovation Center Announcement Letter and Mission Statement 91
B. Interview Coding Framework... 97
C. Members of the 2015 Justice Innovation Center Advisory Panel........................... 99
D. Advisory Panel Read-Ahead Material.. 101
E. Tier 1 Needs from the Justice Innovation Center Advisory Panel, by Agency Type 119

Bibliography.. 127

Figures and Tables

Figures

1.1.	U.S. Rural and Border Counties	2
5.1.	Needs, by Issue Area	76
5.2.	Tier 1 Needs, by Issue Area	76

Tables

S.1.	Summary of Key Findings	xxii
1.1.	The Justice Innovation Center's 12 Communities of Practice	10
2.1.	Definitions for Communities of Practice	16
2.2.	Definitions of Small and Tribal Agencies	17
2.3.	Number of Interviews Conducted, by Agency and Sector Type	21
2.4.	Interviews Conducted, by State and Sector Type	22
2.5.	Agencies, by Community of Practice: Law Enforcement and Courts	24
2.6.	Agencies, by Community of Practice: Institutional and Community Corrections	24
2.7.	SRTB Advisory Panel Participants	27
3.1.	Percentage of State and Local Law Enforcement Agencies Using Various Technologies	32
5.1.	Summary of Major Reported Shortfalls Covered in Panel Read-Ahead Materials	75
6.1.	Summary of Key Findings	89
B.1.	Interview Coding Framework	97
C.1.	Advisory Panel Members and Their Affiliations	99
E.1.	Agency Types, Issues, and Recommendations: Geography	119
E.2.	Agency Types, Issues, and Recommendations: Information Technology	120
E.3.	Agency Types, Issues, and Recommendations: Operations	122
E.4.	Agency Types, Issues, and Recommendations: Funding and Resourcing	125

Summary

Technology is important to improving the effectiveness, efficiency, and safety of the criminal justice system. The development of new technologies and new approaches for applying these technologies has been and will likely continue to be an important catalyst for improvement in law enforcement, corrections, and the courts. However, use of technology in these sectors can be challenging, particularly for agencies located in small, rural, tribal, and border (SRTB) areas. SRTB justice systems account for three-quarters of all criminal justice agencies nationwide. Because these agencies are so widespread and have relatively few employees, they lack a centralized voice to influence the development of technologies and other solutions. To date, relatively little research has examined the needs of such agencies.

The National Institute of Justice created the Justice Innovation Center (JIC) to provide current, rigorous, and actionable information on technology needs and priorities specific to SRTB agencies. The purpose of JIC is to gather information on the challenges that SRTB agencies face, identify relevant technology solutions that can address those challenges, and assess these technology solutions as they are implemented in real-world situations. These activities will provide actionable guidance to SRTB agencies for prioritizing, planning, and implementing technology. The JIC research team consists of staff from RAND Justice Policy and the Arizona State University Center for Violence Prevention and Community Safety.

Although there is no precise or commonly accepted definition of SRTB law enforcement, corrections, and court systems, our project defines SRTB agencies as follows:

- law enforcement departments with fewer than 25 officers
- courts serving districts with fewer than 100,000 inhabitants
- jails with average daily populations of fewer than 50
- probation and parole agencies with fewer than 25 full-time officers
- any of these agency types located in counties with fewer than 2,500 residents and non-adjacent to a metro area
- any of these agency types located in a border county.

In general, the term *rural* refers to areas with low-density populations at a specified distance from large cities and metropolitan areas. We define a rural law enforcement agency as one located in a county characterized as having either (1) a census-defined rural population; (2) a census-defined urban population of less than 2,500; or (3) a census-defined urban population of between 2,500 and 19,999 in an area not adjacent to a metropolitan area. We define a border agency by proximity (100 miles) to either the Canadian or Mexican border. For this study, we define a tribal agency as a local law enforcement agency independently operated by one or a confederation of the 566 federally recognized tribes.

JIC's initial efforts were spent gathering information on the challenges that SRTB agencies face and prioritizing the needs of criminal justice agencies in SRTB contexts. JIC staff also worked to identify relevant technology solutions that can address the highest-priority needs, an effort that will continue throughout the life of JIC. These efforts were carried out using a variety of methods, including a literature review, in-depth interviews with nearly 150 practitioners and topical experts, and focused discussions with an advisory panel of expert practitioners.

Key Findings: Literature Review

General Challenges That SRTB Agencies Face

JIC's in-depth literature review identified some general challenges that SRTB agencies face and some differences across sectors (SRTB) and justice agency types (law enforcement, courts, and corrections). SRTB agencies face many public safety issues similar to those found in urban areas, including crimes imported from urban areas, such as gang activity and drug use (Weisheit, Falcone, and Wells, 1994). SRTB agencies also face unique challenges in addressing crime, including long travel distances for law enforcement response, community supervision appointments, and court attendance; low case volumes, which incentivize multicounty or regionalized systems to deal with limited local budgets and increase efficiency; limited staff and infrastructure, which require staff to perform multiple roles; and lack of local community-based experts and resources (Vetter and Clark, 2013). Moreover, because administrative staff sizes and budgets in SRTB agencies are typically small, training can be cost-prohibitive because of high travel expenses or the time required away from duty. As a consequence of geographic spread and relatively few employees, SRTB agencies have difficulty acquiring grant funding and training for new technologies and lack a centralized voice to affect policy and technological needs assessment (Sale, 2010).

Some issues are especially challenging for certain sectors. SRTB courts can face additional challenges stemming from a lack of programs and services to address mental health, substance use, indigent defense, and self-represented litigants, as well as the increasing need to address language barriers throughout court proceedings (Nugent-Borakove, Mahoney, and Whitcomb, 2011). Border agencies face special challenges related to drug trafficking and human trafficking concerns; the need to collaborate with federal agencies performing border patrol activities heightens the complexity of enforcement. There is little published literature on the tribal justice system, although we do know that, despite declining crime rates nationwide, the most-recent evidence suggests that crime (e.g., violent crime, driving under the influence) disproportionately affects tribal areas (Perry, 2004).

Little Existing Research on the Technology Needs of SRTB Agencies

The literature review highlighted the limited amount of existing research on the technology needs of SRTB agencies. Most studies of technology in criminal justice agencies have focused on larger agencies and those in urban settings, some of which are relevant to SRTB agencies as well. Research conducted as part of RAND's Priority Criminal Justice Needs Initiative identified the three highest-priority technology needs within the law enforcement community generally: improving knowledge of technology and practice within the community; sharing and using law enforcement information both within and across agencies; and researching, developing, testing, and evaluating information technology (IT) for systemwide improvements (Hollywood, Boon, et al., 2015).

Existing research on SRTB technologies focuses mainly on their use in law enforcement agencies. For example, the Bureau of Justice Statistics' Law Enforcement Management and Administrative Statistics survey (Bureau of Justice Statistics, 2015a) found that, in general, non-SRTB agencies have the greatest access to or use of technology, followed, respectively, by border, small, rural, and tribal areas.

Lessons from some studies of large or urban agencies are relevant to SRTB agencies' decisions about technologies. Current research at RAND, conducted under the Priority Criminal Justice Needs Initiative, identified high-priority areas for community corrections agencies (Jackson et al., 2015). Many of these needs were related to improving "fragmented" systems of information-sharing among siloed agencies, improving data extraction and usability by a range of users, and improving training and training materials for handling offenders with mental health issues. The study also identified high-priority needs for institutional corrections agencies (jails and prisons), including issues related to detecting contraband, managing inmate use of technology, and managing and monitoring the enormous amounts of data that the technology systems used in the facilities generate.

Need for Better Data on SRTB Agencies and Technologies

The literature review highlighted the dearth of information on the needs of SRTB agencies and the technologies. We found, furthermore, that there is even less information available measuring the *effects* of technologies for criminal justice agencies, including the value that technology might provide through increased effectiveness, more-efficient use of resources, and other outcomes, such as officer and employee safety.

Key Findings: Interviews

To gain a better understanding of the technology needs of SRTB agencies, we conducted nearly 150 interviews with representatives from SRTB agencies, including personnel from law enforcement, courts, and institutional and community corrections (the latter being probation and parole agencies). Our interviews included representatives from 29 border agencies and 25 interviews with representatives from tribal agencies.

Taken together, the SRTB interviews revealed a broad range of operational challenges that clustered around a few central themes, including challenges and barriers related to geography, types of crime common in these communities, funding, personnel, infrastructure, IT (both in capacity and implementation), data-sharing and communication, and legal and policy issues. In this section, we discuss major crosscutting themes. We present them in no particular order.

Geographic Challenges

Many respondents across SRTB agencies mentioned challenges associated with geographic isolation and challenging topography. Geographic isolation can lead to staffing problems because of transportation times for police officers responding to calls, community corrections and court officers to conduct community supervision, and jail staff to transport inmates for external health services and court appearances. Specific challenges for courts and corrections facilities included difficulty in getting jurors and defendants to courthouses on trial dates because of long travel times and a lack of transportation resources in the community. Geography also makes the challenge of combating drug cartels with superior technological capabilities more difficult.

Funding Challenges

Interview respondents universally mentioned funding shortfalls, and those shortfalls colored many of the other challenges that the communities of practice face.[1] Many respondents reported that their budgets have been decreasing over time. Funding often lags behind need, leading to low wages and slow adoption of new technology and, in turn, difficulties attracting and retaining qualified staff. Respondents throughout the spectrum of participating agencies reported that small budgets imperiled recruiting, retention, and training.

Other problems followed from a lack of funding. The rarity of discretionary funding for technology acquisitions contributed to slow adoption of potential technology solutions to known needs. Moreover, many small and rural agencies reported struggling to meet requirements from new policies or mandates because of limited resources. For example, new camera requirements in many jurisdictions for both police and corrections agencies can strain existing resources and facilities. Funding constraints among SRTB agencies also meant that many agencies could not take advantage of economies of scale for technology acquisition. As a result, SRTB agencies often struggle to take on technology projects requiring large capital investments.

Social Service Provision Challenges

Respondents noted a variety of technology-related challenges that affect service provision, especially related to mental health and substance-use treatment; these issues appear to affect people who might come into contact with any element of the justice system. Court respondents noted that small courts cannot benefit from the specialization of judges and courts (e.g., problem-solving courts) because larger, well-resourced establishments are typically necessary to justify the investment for such specializations. Institutional corrections personnel reported interest in telemedicine and remote treatment services for inmates but noted the large start-up costs, financial and otherwise, that can hinder implementation. Community corrections personnel described difficulties in getting their clients into services, whether because of the high program costs paid by the probationer, the lack of programs available in the local area, or the lack of openings in existing programs. Language barriers between practitioners and others in the criminal justice system were also common across sectors, especially in border areas. Finally, courts and corrections agencies indicated a lack of transportation, job training, substance-use treatment, and mental health and medical services for clients.

Personnel Challenges

Budget challenges and geographic isolation can contribute to problems with staff recruitment, retention, and training for SRTB agencies. Constrained budgets leave little money available for hiring or training, and salaries at SRTB agencies are often not competitive with those at

---

[1] Throughout, we define an agency type based on its functions (law enforcement, courts, or corrections) and sector type on its size or location (small, rural, tribal, or border). We use *community of practice* to mean a combination of the two (e.g., small law enforcement agencies are a community of practice).

larger agencies. A lack of training opportunities can lead to additional challenges for the use of technology, which often requires extensive training for effective use. Retention was also a challenge for many SRTB agencies because of the more-competitive salaries and benefits offered at larger agencies. Agencies also face difficulty in getting support in specialized areas, including IT support and translation services.

Communications and Information-Technology Management Challenges
Many agencies seeking to implement technology faced challenges in IT management. These challenges included antiquated case-management systems (CMSs) and jail-management systems (JMSs), a lack of standardized electronic data-collection technology across SRTB agencies, and management and analysis software problems caused by poor interface design, needless complexity, or a lack of desired capabilities. All sectors provided evidence of insufficient IT capacity among personnel.

Some of the technological deficits noted among SRTB law enforcement agencies included a lack of laptop computers and cameras in patrol cars, license-plate readers, and radios. Among corrections facilities, interviewees reported insufficient bandwidth to run many systems, including video visitation, telemedicine, and remote arraignment. Community corrections respondents highlighted struggles with technologies that are considered standard in larger or urban agencies. Border criminal justice agencies reported challenges stemming from their locations near international borders, which they frequently linked to difficulties with operating technology.

Data-Sharing and Interoperability Challenges
Challenges related to information-sharing and interoperability across agencies were also widely reported. Many respondents reported that being unable to access the data they needed to do their jobs well impeded their ability to carry out their duties. For example, several court respondents noted that a lack of data-sharing infrastructure prevents judges and qualified staff from electronic access to important databases, while a lack of web infrastructure at some courts makes it difficult for the public to access public records. Corrections agencies reported mixed success with their JMSs and record-management systems. Other agencies reported a lack of remote access to data, as well as frustrations with state systems, that kept counties in individual silos, preventing an easy flow of information between different state offices.

Border agencies face special problems with data-sharing and interoperability because of the need for collaboration on some issues (e.g., immigration) with federal law enforcement agencies or with agencies operating across international borders. Such needs exacerbate shortcomings in data-sharing and interoperability with other local and state agencies, including local law enforcement, and federal agencies.

Crime-Specific Challenges
SRTB agencies often have to deal with particular types of crime, as interviewees from law enforcement agencies and from border agencies highlighted. These included various issues specific to individual communities, including problems related to alcohol and drug use, domestic violence, and drug and human trafficking. Other crime-related issues that respondents noted include the prevalence of casinos in some SRTB areas, which some interviewees felt created and attracted a significant amount of crime.

Infrastructure Challenges

Limitations in facility and technological infrastructure were common in SRTB agencies, particularly courts and institutional corrections facilities. Respondents described many facilities as old and in need of repairs and renovations to be suitable for technology upgrades. Some courts, particularly tribal courts, faced challenges with the structural adequacy of their buildings. Other facilities were so antiquated that some technology acquisition would make sense only when a new facility is built. Facility shortcomings often created security concerns for employees, with some respondents reporting unreliable utilities, even including electricity and Internet connectivity. Several court personnel reported that they lack basic security measures, such as barriers between defendants and judges, and screening technologies, such as metal detectors.

Resistance to Technological Change

For many SRTB agencies, the rarity of discretionary funding available for technology acquisitions, along with lack of technological infrastructure and training opportunities, contributes to the slow adoption of potential technology solutions, even for known needs.

We noted these themes in particular among court and corrections respondents. For example, a few respondents described courts as slow to adopt innovations because of a culture of distrust of technology, and several respondents noted that the heavily used practice of manually filing forms causes confusion and delay among the growing number of pro se (i.e., self-represented) litigants. Many community corrections agencies focused less on new and sophisticated technologies and more on improvements to basic technologies, such as phone systems, Internet access, and computers. Interviewees had very mixed responses regarding smartphones: Some agencies issued them to all officers, others only to supervisors or officers with high-risk clients; others were in the process of piloting smartphones; and still others either did not feel that they were necessary or did not have interest from staff or the budget to support an upgrade to smartphones.

Legal and Policy Barriers

Challenges created by external policies and mandates were also common across the sectors. Examples from law enforcement and corrections included mandates to increase camera use without sufficient funding for the implementing agencies. Other policies forbade technology options with potential benefits, exemplified by document e-filing, which is prohibited in many courts despite respondents' enthusiasm about the potential reduction in administrative costs. Court respondents cited legal and policy barriers as significant obstacles in the implementation of technologies and best practices. One respondent noted that some CMSs can be difficult to adapt to statutory and policy changes.

Across 25 interviews with representatives of tribal agencies, as well as interviews with representatives of agencies bordering tribal nations and the Bureau of Indian Affairs (BIA), we identified a range of operational needs that might be unique to, or more common among, tribal justice agencies. In particular, two laws play a significant role in determining differences across tribal agencies, their authorities, resources, and operational requirements. Before the enactment of Public Law 83-280 in 1953, sovereign tribal nations and the federal government had concurrent jurisdiction over criminal and civil matters in tribal lands, similar to the jurisdictional arrangements that states have with the federal government. With Pub. L. 83-280, Congress shifted federal jurisdiction to state governments in six states: California, Minnesota, Nebraska, Oregon, and Wisconsin (and, when it became a state, Alaska). Since then, other states have assumed full or partial concurrent jurisdiction in tribal lands under Pub. L. 83-280. A second law of note is the Indian Self-Determination and Education Assistance Act of 1975 (Pub. L. 93-638), which provides the mechanism whereby tribes could take over the administration of services that the federal government had, until then, provided. Tribes with Pub. L. 93-638 agencies administer their own law enforcement, courts, and corrections systems under so-called self-determination contracts with the BIA. Later, a second funding mechanism called a BIA compact was created that provided some tribes with block-grant funding that could be allocated across a range of approved activities, including support for justice system agencies.

Interviewees described several challenges stemming from issues related to governance and tribal sovereignty. For instance, one tribal police chief explained that there was no existing tribal domestic abuse law his agency could enforce. Instead, prosecutable cases were handled as sexual assaults or batteries. One chief mentioned that the tribal community was guarded about its sovereignty and that its members did not want to share a lot of information with outsiders and did not necessarily trust the police department. One clerk explained that her tribe, which has a service area but not a reservation, is in a Pub. L. 83-280 state, meaning that the state has jurisdiction over criminal prohibitory issues but not enforcement of orders. Another clerk noted that, because her reservation encompasses four different tribes across two states, "coordination and jurisdictional issues . . . are constantly coming up."

Key Findings: Expert Panel

We used the information gathered via the interviews to develop the agenda and discussion points for a JIC advisory panel meeting. JIC's advisory panel, which consisted of experts on and practitioners from SRTB criminal justice agencies, was tasked with identifying and prioritizing a set of science- and technology-related *needs*. Here, we define a need to be a *requirement for innovation to address a problem (or opportunity) that SRTB agencies face*. The *need* is related in some way to science and technology but might include technological development, development of guidance and educational materials, development of training, development and dissemination of new business practices, and development and dissemination of model policies or even legislation.

The JIC advisory panel convened on December 7–8, 2015, in RAND's Arlington, Virginia, office. Day 1 of the meeting was devoted first to reviewing the key issues that SRTB agencies face and initial ideas on solutions. Day 2 focused on prioritizing needs for each agency type. During discussions, panels often held discussions in four breakout groups (one each on law enforcement, courts, institutional corrections, and community corrections).

Overview of Panel Results

Two general points of interest cut across all the breakout groups' discussions.

The first is that the majority of issues and needs discussed were not SRTB-specific; they apply to larger agencies as well. The results were broadly consistent with needs generated through RAND's Priority Criminal Justice Needs Initiative (Hollywood, Boon, et al., 2015). All breakout groups had top-ranking needs related to improving the sharing and use of information, as well as needs for providing educational support to improve agencies' knowledge of technology and how to acquire and use it effectively. Notable exceptions had to do with the needs related to geographic challenges (agencies have to cover long distances; people with key skills and service providers typically are not local) and resourcing challenges (smaller populations meant that SRTB agencies were extremely resource-challenged; small size further reduced agencies' capabilities to acquire technologies and seek grants and other funding). Tribal and border agencies also reported some sector-specific challenges.

The second point of interest is that panelists from all four agency types reported being challenged by some sort of major societal or cultural change. For law enforcement, it was the need to improve relationships with the broader community. For courts, it was the need to support an influx of litigants representing themselves (addressing litigants who did not speak English was a related concern). For institutional corrections, it was the need to provide greatly improved mental health and other treatment services to inmates. Finally, for community corrections, it was the need to focus efforts on rehabilitation and positive behavioral change rather than continued punishment.

The advisory panel identified a range of needs for each agency type. Approximately one-third of all needs were focused on IT (e.g., record-management systems, data-sharing capabilities) and one-third on operations (e.g., daily job requirements). The remaining third was split between funding and geographic issues. Within each agency type, however, different patterns emerged. Law enforcement and institutional corrections needs focused heavily on operations; court needs focused heavily on IT issues; and community corrections needs focused heavily on resourcing issues.

All breakout groups had some top-ranking needs related to improving the sharing and use of information, as well as needs for providing educational support to improve agencies' knowledge of technology and how to acquire and use it effectively.

In the rest of this section, we describe some of the top themes in each sector.

Law Enforcement Themes
Improving Relationships with the Community
The general need to improve relationships and build trust with the community was a very strong theme in the panelists' discussions. The top-rated need was to improve agencies' abilities to work with the media. Another need was to disseminate strategies on how to increase agencies' presence in their communities. Technically oriented needs included developing social media apps that allow community members to provide feedback.

Improving Information-Sharing for Law Enforcement and Leveraging Common Standards
Specific needs called for assessing the feasibility of state and national standards for sharing state- and federal-level data with agencies and developing corresponding "patches" for legacy systems to incorporate those standards. There was also a call to develop models for becoming compliant with the National Incident-Based Reporting System standard for reporting crime data.

Addressing Information-Technology Management Shortfalls
The top-ranking need here was to develop tools to help small agencies better determine their IT needs and match those needs with specific systems and vendors. Other needs addressed IT system management.

Other key needs identified for law enforcement included those related to improving communication infrastructure (e.g., by upgrading to digital equipment), getting a better understanding of body-worn cameras, addressing responses over long distances (one controversial proposal focused on creating a model memorandum of understanding to permit neighboring agencies to respond under specified conditions), and helping agencies build grant-development capabilities through technical assistance (TA) centers.

Court Themes
Addressing the Surge in Pro Se Litigants
Participants saw addressing the influx of pro se litigants, many of whom have no experience with court processes and proceedings, as a major challenge. Specific needs here included developing a Turbo Tax–like app to help litigants complete legal forms, as well as guides, brochures, and videos to explain court procedures to the public. Somewhat related to this theme was a need to expand SRTB courts' capabilities to work with litigants who do not speak English, possibly by creating a rapid certification process for court translators.

Improving Security and Resiliency at SRTB Courts
Several top-ranked needs applied to improving the security and resiliency of SRTB courts through such means as continuity-of-operations planning tools, emergency response training, and evaluations of courthouses' abilities to respond to emergencies.

Improving Information-Technology Infrastructure for Courts in General
There was a strong desire to build up courts' IT infrastructure in general. There was also a high-rated need to assess whether courts could buy systems to leverage new First Responder Network Authority (FirstNet) capabilities. Participants called for assessments of the use of satellite communications in areas not well served by cellular service. Similarly, there was a call to assess using general-purpose, low-cost communication systems (such as voice-over–Internet protocol services) as potential replacements for purchasing customized communication systems.

Other areas of need focused on improving information-sharing for courts (e.g., through model policies and a planning process), improving technology training (e.g., by increasing the use of online training courses), and addressing funding shortfalls (e.g., through technology training and funding of IT-related TA).

Institutional Corrections Themes
Improving the Provision of Mental Health and Other Services to Inmates
The top-rated need from the group was for jails and academies to provide deescalation training. A related need was for jails to set up, train, and use specialized critical incident or crisis intervention teams to respond to inmates in crisis. There was also a call to provide agencies with a mental health assessment tool for incoming inmates that can be used to support decisions regarding care and placement.

Developing Corrections Personnel
The need to improve the capabilities of both agencies and individual employees (and employee retention) was another top theme from the institutional corrections breakout group. Deescalation training was the most requested type of training. The second-rated need was a call for agency heads to better develop line staff and actively involve them in the workings of institutions. There was also a call for agencies to employ American Correctional Association standards.

Improving Information-Sharing, Especially Via Improving Jail-Management Systems

Needs here began with a call for a checklist of mandatory elements that should be in a JMS. There were also calls to develop a program for certifying commercial providers as experts in corrections IT systems and to fund IT experts to provide TA on corrections systems.

Addressing Funding Shortfalls

Panelists noted that agencies have to devote substantial time and resources to seeking funding and conducting procurement. The top need in response was a suggestion that federal grant regulations be modified so agencies could hire part-time employees to administer grants.

Community Corrections Themes
Refocusing on Rehabilitation and Positive Behavioral Change

Participants emphasized a broad-based need for community corrections to transition to focusing on rehabilitation and positive behavioral change, not continued punishment, as their main purpose. Corresponding needs here included a call for the community to redefine the role of officers as a combination of law enforcement and social work and a call for officers to connect clients with people who could provide what they called "role-modeling behavior" in their communities. Participants also asked that performance information from monitoring technologies (Global Positioning System [GPS] tracking, substance-use tests) be used to provide positive reinforcement, not just to punish lapses.

Improving Information-Sharing

Participants described a need for mobile data terminals in vehicles for community corrections officers. The baseline and top-rated need in this area was to develop a tablet or phablet (smartphone and tablet in one) that would include CMS functions and integration, along with access to other key databases and task-specific software.

The second need focused on providing information to community corrections officers, i.e., on building statewide or regional platforms to conduct agency, state, regional, and federal information-sharing.

Addressing Geographic Dispersion

The top-rated needs from the community corrections breakout group all focused on ways to overcome geographic challenges in order to provide rehabilitative services to clients. Two needs focused on leveraging videoconferencing (e.g., for officer–client visits, for mental health and other service providers' sessions). A third called for building relationships with law enforcement so that community corrections and law enforcement officers could reinforce each other's efforts in remote areas.

Addressing Funding Shortfalls

The top-rated need related to resourcing challenges was to provide agencies with support in identifying new grant opportunities and writing corresponding grant proposals. Other needs included helping agencies assess the feasibility and costs of technology upgrades to be funded through grants and asking agencies to develop staffing models that permit the use of support staff (i.e., nonsworn personnel) to support clients' nontherapeutic needs.

Crosscutting Themes

Some needs across the agency types were very similar, cutting across the criminal justice community.

Information-Sharing Between Agency and Other Governmental Systems

This theme represented challenges of sharing data across different agencies and other governmental systems. For example, courts, jails, and law enforcement cannot access one another's data systems, which makes coordinating services challenging.

Assisting Agencies with the Procurement and Management of Information-Technology Systems

The expert panel identified a need for remote training for IT, developing technology mentorship and TA funding, and developing a low-cost way to improve internal IT management capabilities.

Use of Videoconferencing to Overcome Distance Barriers

The panel described a variety of needs related to using videoconferencing tools both for internal (hearings, client meetings) and external uses (treatment sessions).

Assistance in Obtaining Grants

Agencies need better education about available funding opportunities, TA to help agencies write grant applications, and ways to help them reduce the time spent seeking and applying for funding.

Using Nonstandard Personnel

Needs covered assessing the use of so-called nonstandard (i.e., nonsworn) personnel to carry out certain functions at lower cost, including IT system administration, grant administration, and handling clients' nontherapeutic needs.

Conclusions

JIC's first year of research helped to fill in some gaps in understanding concerning the technology challenges and needs that SRTB agencies face. In Table S.1, we briefly recap the key findings from each of our activities.

One important lesson that comes from looking across the findings from the first year is that, although SRTB agencies report many common themes, each agency type also faces some issues that are prominent within that type of agency, such as the need for law enforcement agencies to improve relationships with communities or the need for courts to address the surge in pro se litigants. This means that technology solutions must not only be capable of addressing the broad challenges faced by SRTB agencies (e.g., the need to cover long distances, lack of funding)—many of which agencies of all sizes share—but must also address the specific needs of each agency type and of individual agencies.

Another important lesson from the first year is that better data are needed concerning the effectiveness of technologies to address the challenges that SRTB agencies face. JIC's literature review found only a few studies that examine the effectiveness or utility of technology in general, and very few rigorous evaluations of technology have been conducted with SRTB

Table S.1
Summary of Key Findings

Method	Key Finding
Literature review	• Small, rural, tribal, and border (SRTB) agencies face many issues similar to those in urban areas, including gang activity and drug use. • SRTB agencies also face unique challenges, including long distances, low case volumes, limited staff and infrastructure, and lack of community-based experts and resources. • SRTB agencies have difficulty acquiring grant funding and training for new technologies and lack a centralized voice to affect policy and technological need assessment. • Some issues are especially challenging for certain sectors, such as the court's lack of mental health and substance-use programs, and border agencies' need to address drug and human trafficking. • There is little existing research on the technology needs of SRTB agencies.
Interviews	• Funding shortfalls were universally mentioned and colored many of the other challenges faced by our 12 communities of practice. • The rarity of discretionary funding contributed to slow adoption of technology solutions and made it difficult for agencies to take advantage of economies of scale. • Geographic challenges ranged from difficulties in getting jurors and defendants to courthouses on trial dates to long transportation times for law enforcement officers responding to calls. • Information-technology (IT) challenges included antiquated case-management systems (CMSs) and jail-management systems (JMSs) and SRTB agencies' lack of technology compared with their urban counterparts. • Facility and physical infrastructure limitations were common. • Difficulties with information-sharing and interoperability across agencies were widely reported. • Language barriers between practitioners and others in the criminal justice system were common across sectors. • Tribal agencies face some unique challenges related to governance and tribal sovereignty.
Expert panel	• The majority of issues and needs discussed were not SRTB-specific but apply to larger agencies as well. • Panelists from all four agency types reported being challenged by some sort of major societal or cultural change. • Key law enforcement needs focused on improving relationships with the community, improving information-sharing, and addressing IT shortfalls. • Key court needs focused on addressing the surge in pro se litigants, improving security and resiliency at SRTB courts, and improving IT infrastructure. • Key institutional corrections needs focused on improving the provision of mental health and other services to inmates, developing the capabilities of corrections personnel, improving information-sharing through better JMSs, and addressing funding shortfalls. • Key community corrections needs focused on transitioning to a focus on rehabilitation and positive behavioral change, improving information-sharing, addressing geographic dispersion, and addressing funding shortfalls. • Key needs across agencies included better information-sharing, assistance with procurement and management of IT systems, greater use of videoconferencing to overcome distance barriers, assistance in obtaining grants, and using nonstandard personnel to carry out certain functions at lower cost.

agencies. SRTB agencies need better guidance on how to set clear and defensible technology priorities.

JIC's future research will take steps to address these issues. This subsequent work will focus on assessing technology solutions in specific, real-world SRTB contexts, helping to build the knowledge base on technology effectiveness in the SRTB context. Also, it will help build up a set of tools and resources (e.g., website, searchable database) that agencies can use to develop customized technological solutions to address both the common issues that many agencies face, as well as individual agencies' specific needs.

Plans for Future Research

We will use the results of the advisory panel meeting to identify technologies that JIC will evaluate in its subsequent research. These evaluations will examine the costs and benefits associated with agencies' adoption of new technologies, considering acquisition costs, staff time, training requirements, system operation and maintenance costs, safety improvements, and performance outcomes. Once technologies have been selected for evaluation, JIC staff will recruit agencies willing to participate in research implementations of the technology. These evaluation projects will be designed to provide information that can be used to guide other agencies' decisions on what technology to use, what pitfalls to avoid, and what improvements they might get from the technology.

In the future, JIC will also continue to update the literature review as new sources are published or identified. JIC researchers are also contributing to a wiki that is under development at RAND; the wiki will act as a clearinghouse for information useful to practitioners and others interested in learning more about technology in criminal justice agencies. Finally, JIC will continue to disseminate information about its work through a redesigned website, as well as conferences of relevance to SRTB criminal justice agencies.

Acknowledgments

The study reported here was made possible through a grant from the National Institute of Justice. We are grateful to our project officer, Michael O'Shea, for his assistance throughout the project.

This project was a partnership between RAND and the Arizona State University Center for Violence Prevention and Community Safety. We are grateful to all of the staff who participated in the development and implementation of the first year of this project. This includes RAND staff Eyal Aharoni, Samantha Cherney, Cordaye Ogletree, Robert Stewart, Quinton Stroud, and Sarah Weilant and Arizona State University staff Marie Griffin, John Hepburn, Charles Katz, Weston Morrow, and Michael White. We also thank RAND staff Katherine Mariska and Natalie Kauppi for their administrative work on the Justice Innovation Center and Liisa Ecola for helping coordinate the response to reviewer comments.

Our quality assurance reviewers, Lois Davis of RAND and Lisa Growette Bostaph of Boise State University, made valuable contributions to the clarity and comprehensiveness of this report. We thank them for making the time to provide such perceptive reviews.

We would also like to thank all the people who took time to participate in the interviews and advisory panel. Their insights were invaluable.

Abbreviations

ABA	American Bar Association
ACA	American Correctional Association
AJA	American Jail Association
BIA	Bureau of Indian Affairs
BJA	Bureau of Justice Assistance
BJS	Bureau of Justice Statistics
BWC	body-worn camera
CMS	case-management system
CSLLEA	Census of State and Local Law Enforcement Agencies
DHS	U.S. Department of Homeland Security
DOJ	U.S. Department of Justice
DUI	driving under the influence
FBI	Federal Bureau of Investigation
FCC	Federal Communications Commission
GPS	Global Positioning System
HIV	human immunodeficiency virus
IACP	International Association of Chiefs of Police
IT	information technology
JIC	Justice Innovation Center
JMS	jail-management system
LECTAC	Law Enforcement and Corrections Technology Advisory Council
LPR	license-plate reader
MOU	memorandum of understanding

NAICJA	National American Indian Court Judges Association
NCSC	National Center for State Courts
NIBRS	National Incident-Based Reporting System
NIJ	National Institute of Justice
NLECTC	National Law Enforcement and Corrections Technology Center
OIG	Office of the Inspector General, U.S. Department of Justice
OJP	Office of Justice Programs
PREA	Prison Rape Elimination Act
PSI	presentencing investigation
RMS	record-management system
SRTB	small, rural, tribal, and border
TA	technical assistance
USDA	U.S. Department of Agriculture

CHAPTER ONE

Introduction

Technology is important to improving the effectiveness, efficiency, and safety of the criminal justice system. The development of new technologies and new approaches for applying these technologies has been and will likely continue to be an important catalyst for improvement in law enforcement, corrections agencies (both institutional corrections, such as jails and prisons, and community corrections, such as probation and parole agencies), and the courts. However, use of technology in these sectors can be challenging. Given the large number of police departments—including about 18,000 law enforcement agencies alone—corrections agencies, and court systems, the market for criminal justice technologies is fragmented, and investments are diffused across multiple sectors. For a variety of reasons that we discuss in this report, this challenge is felt acutely in small, rural, tribal, and border (SRTB) areas.

The National Institute of Justice (NIJ) created the Justice Innovation Center (JIC) to provide current, rigorous, and actionable information on technology needs and priorities specific to SRTB agencies, focusing on 12 communities of practice spanning the areas of law enforcement, courts, and institutional and community corrections.

JIC's work supports NIJ's investments in justice system technologies and informs the wider communities of technology providers and adopters. JIC also provides NIJ with guidance on which technologies can be cost-effectively employed in SRTB settings and where valuable technical assistance (TA) resources would yield the greatest benefits. The JIC research team consists of staff from RAND Justice Policy and the Arizona State University Center for Violence Prevention and Community Safety.

This report reviews the work of JIC, provides findings for its main data collection effort—interviews with representatives from agencies across the criminal justice spectrum in SRTB areas—and looks ahead to upcoming research, evaluation, and practitioner support activities planned for future years. The remainder of this introduction provides background on the types of agencies and contexts with which JIC is concerned (small, rural, tribal, and border) and outlines the activities undertaken thus far.

How Are SRTB Agencies Different?

According to the 2010 U.S. census, about 20 percent of U.S. residents, or 59 million people, live in rural areas, and nearly three-quarters (73.6 percent) of those people live in the South or

Midwest regions.[1] The percentage of U.S. residents living in rural areas changed little between 2000 and 2010, declining only about 1.6 percentage points. With approximately one in five U.S. residents living in rural areas, a significant level of criminal justice services need to be delivered in challenging geographic landscapes, often with limited resources, and without some technologies considered standard in urban areas. These conditions can create barriers to residents' access to public safety and justice. Figure 1.1 shows the distribution of counties classified as rural using the U.S. Department of Agriculture (USDA) rural–urban continuum for 2013 and border counties (those within 100 miles of either international border).

More than one-quarter of the 1.7 million members of the Native American and Alaska Native populations live in rural areas. Rural areas are also experiencing growth in their Hispanic populations. Increased rural diversity adds to the challenges that criminal justice agencies face in these areas. Further, the political climate surrounding immigration policies that affect police, courts, and corrections agencies and extensive coordination with federal authori-

Figure 1.1
U.S. Rural and Border Counties

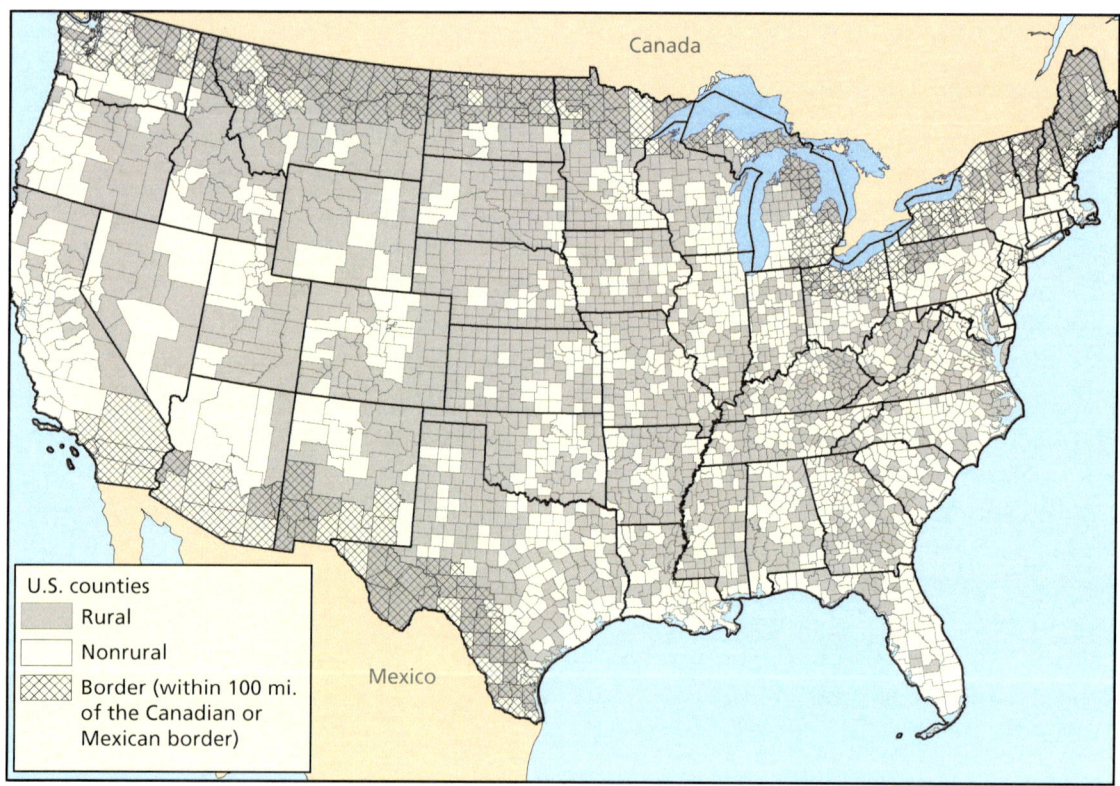

DATA SOURCE: Economic Research Service, 2013.
NOTE: We do not show Hawaii or Alaska. All counties in Hawaii are nonrural. Only four counties in Alaska are nonrural (those in the metro areas of Anchorage, Fairbanks, and Juneau). A rural county is one classified as code 6, 7, 8, or 9 on the rural–urban classification scale.
RAND RR1479-1.1

---

[1] The U.S. Census Bureau defines urbanized areas as having 50,000 or more people and rural areas as encompassing all population, housing, and territory not included within an urban area.

ties responsible for enforcing immigration laws adds a layer of complexity to the administration of justice in border areas, which are often also rural or small.

Relatively few studies have examined SRTB criminal justice agencies. Here, we describe some of the limited research that has contrasted small, rural, and tribal agencies with larger, urban, or nontribal agencies or explored the unique needs and challenges that these agencies face. In general, SRTB agencies face a range of challenges related to geography, demographics, legal status, and resources. Few studies have examined these agencies in the context of border issues.

Small and Rural Agencies

Law Enforcement

In general, SRTB law enforcement agencies look very different from their non-SRTB counterparts. Data from the 2008 Census of State and Local Law Enforcement Agencies (CSLLEA) show that small, rural, and tribal agencies tend to have lower operating budgets and personnel than border and non-SRTB agencies (Reaves, 2011).

Small law enforcement agencies have been described as more community-oriented and informal and less professional or militaristic than their larger, urban counterparts, although the empirical evidence on this is mixed. For example, although police in smaller agencies are reported to maintain closer ties to citizens, as well as to be better acquainted with local citizens (Liederbach and Frank, 2003), police in larger agencies engage in a wider variety of community policing activities (Maguire et al., 1997). Other studies have found that per capita expenditures (per officer and resident) are lower and clearance rates are higher in small agencies, implying greater effectiveness and efficiency than in urban jurisdictions (Falcone, Wells, and Weisheit, 2002).

The work activities of police officers are similar across different-sized departments; for example, Liederbach and Frank found that police officers in small agencies, like their urban counterparts, engage in such tasks as routine patrol and driving, report-writing and other administrative tasks, and personal activities (Liederbach and Frank, 2003). However, officers in small departments tend to deal with a greater variety of crime types and a wider array of issues and are more likely to be generalists than to specialize (Liederbach and Frank, 2003). In addition, public safety concerns vary, with smaller departments having less concern for violence and gang activity; for example, Kuhns, Maguire, and Cox, 2007, shows that concerns regarding alcohol offenses, disorderly conduct, vandalism, traffic offenses, wildlife and agricultural crime, and domestic violence are inversely related to agency size. Despite these generalities, there is considerable variation across small-town and rural agencies.

Courts

Small and rural courts face unique issues as well. The so-called court community context provides a framework for understanding differences in the administration of justice in different areas. Ulmer and Bradley, 2006, p. 641, citing Eisenstein, Flemming, and Nardulli, 1988, describes courts as "distinct, localized social worlds with their own relationship networks, organizational culture, political arrangements, and the like." For small and rural courts, which typically have fewer staff members and thus often require staff to take on more-generalized roles, Wilcox et al., 2008, p. 70, suggests that the administration of justice relies "more frequently upon strong connections among court workers for informal processing." Urban courts, on the other hand, "have a higher degree of formality and bureaucratization than those in rural areas" (Wilcox et al., 2008, p. 70). Indeed, as reported in Logan, Shannon, and Walker, 2005, on protective orders, the authors found that women in rural areas believed that people with informal connections to court workers or law enforcement were more likely to have their request-for-protection orders taken seriously.

Many existing studies of SRTB courts have focused on the juvenile justice system, which is distinct from the adult justice system. Several studies have documented variation between urban and rural areas in case administration, processing, and sentencing, particularly with regard to the juvenile justice system. As Feld, 1991, reports on juvenile case processing, the author found that, relative to urban areas, courts in rural areas made fewer felony referrals and had lower detention rates. Feld also suggested that "informal community alternatives" and more-lenient sentences were more common in rural areas. Sims, 1988, also reports more-informal handling of juvenile cases by police in the field but more referrals to the formal juvenile justice system once the decision was made to arrest a juvenile: In other words, once a youth entered the justice system, he or she was *less* likely to be given alternatives to formal processing than in urban areas. Maupin and Bond-Maupin, 1999, offers additional insight into this trend, suggesting that it is driven by a lack of community alternatives in rural places; one judge interviewed for the Maupin and Bond-Maupin study suggested "emergency shelter care, long-term shelter care, and electronic monitoring" as possible alternatives that could help a juvenile avoid detention (p. 19).

Rodriguez, 2008, considers the effects of rurality on juvenile justice for Native Americans and those in border areas. Rodriguez's analysis of court processes in Arizona found that Native Americans were "less likely to be referred for informal processing, more likely to be detained, and more likely to have a petition filed than Whites," all of which could be in part due to the lack of alternatives to detention available through tribal court systems (p. 26). Rodriguez suggests that, in border areas, her findings that Hispanics were more likely to face formal processing and more likely to be detained than whites could be due to the "political climate surrounding illegal immigration and border security" (p. 27).

The studies cited above, many of which focused on the juvenile justice system, indicate that small and rural courts have unique contexts, processes, and outcomes, different from those in urban areas, and that their subsequent needs—for technology or otherwise—are likely to be unique as well.

Institutional Corrections

According to the Bureau of Justice Statistics (BJS), in 2006, nearly 40 percent of jails in the United States were small, holding fewer than 50 inmates (Stephan and Walsh, 2011). These small jails were responsible for 6 percent of all jail admissions in 2014 (Minton and Zeng, 2015). Relative to larger jails (housing 250 or more inmates), small jails had more white inmates (65 percent versus 36 percent), had smaller Hispanic populations (8 percent versus 16 percent), and housed fewer noncitizens (4 percent versus 5 percent)—these figures reflect overall lower levels of diversity in rural areas (where small jails tend to be located) than in urban areas (BJS, 2015b).

Using data from *Census of Jails, 1999* (Stephan, 2001), Applegate and Sitren, 2008, reports that rural jails tended to be older, less crowded, and staffed by a greater proportion of women and to have lower ratios of inmates to corrections officers and corrections officers to administrators and lower rates of assault and communicable diseases, such as tuberculosis and human immunodeficiency virus (HIV) (although the authors attribute this to lower rates of testing).

Small jails, however, experience a much greater rate of inmate turnover than larger facilities do: In 2014, small jails had a turnover rate of approximately 104 percent, while large-jail jurisdictions (with 1,000 or more inmates) had a turnover rate of only 49 percent (Minton and Zeng, 2015). One reason for higher turnover rates in small jails is that a lack of support services (and available alternatives to detention) in small and rural communities can lead to a reliance on jails for handling pretrial and postconviction criminal justice matters (Mays and Thompson, 1988).

Small jails, then, typically handle many more admissions and discharges relative to their overall capacity than larger jails do and are likely to house inmates with problems that, in urban areas, might be handled through community services. And small jails typically offer fewer services, such as on-site medical personnel, work opportunities, education programs, and counseling and rehabilitative programming, than urban jails do. High turnover and inmates with greater needs require more administrative work by staff and challenge these jails' ability to provide adequate and timely services to people who are incarcerated for short periods.

Dealing with inmates' mental health issues is a challenge for jails of all sizes but is especially acute for small and rural jails, for reasons mentioned above. Limited resources, limited training opportunities for staff, and limited facility space—all common in small and rural jails—further contribute to an overall lack of sufficient mental health support services for inmates there (Race et al., 2010).

Community Corrections

Probation is the most frequent sanction utilized in sentencing, but there is little research examining the unique challenges associated with probation in small and rural agencies or comparing small and rural agencies with urban departments. Many reentry challenges, such as lack of housing and homelessness, difficulty finding employment and jobs, and need for substance use and mental health treatment, can be exacerbated in rural areas, which lack available treatment providers and services. These areas also often face more-limited employment opportunities than urban areas do (Wodahl, 2006). Probation officers in smaller agencies tend to have larger caseloads and to lack resources, such as local training opportunities (Thomson and Fogel, 1981).

Several differences between probationers in rural areas and those in urban areas can influence the needs and challenges small and rural probation departments experience. Probationers in rural areas tend to be white and younger and to belong to middle income categories. Probationers in rural areas are also more likely to have committed misdemeanors than felonies, victimized an acquaintance or friend (rather than a family member or stranger), and experienced fewer arrests and convictions but have had more probation sentences than their urban counterparts. In addition, probation staff perceive them to be in greater need of drug and alcohol treatment and to report higher overall prevalence of substance use, but not illegal drug use. Rural probationers are more likely to have been sentenced for driving under the influence (DUI), whereas urban probationers are more likely to have been sentenced for drug-related violations, a finding that is consistent with arrest patterns in rural and urban areas. Rural probationers are also more likely to be required to pay fees and restitution, to be court-ordered to treatment, and to wear electronic monitors as a condition of their probation (Ellsworth and Weisheit, 1997; Olson, Weisheit, and Ellsworth, 2001). However, it is important to note that this body of work is more than a decade old and might not reflect current trends.

Tribal Agencies

Law Enforcement

Research has shown that rates of violence and substance use are higher in tribal areas, but there is a dearth of evidence on the operations of tribal criminal justice agencies. Tribal law enforcement agencies are likely to face many of the same challenges as small and rural agencies: Most are small, employing fewer than ten sworn officers; are located in nonmetropolitan areas serving jurisdictions with fewer than 20,000 residents; and engage in primary police functions that are similar to those of comparably sized agencies (Wells and Falcone, 2008).

On the other hand, tribal police tend to be less involved in criminal investigations than nontribal agencies and more involved in providing supplemental policing services that reflect the unique contexts in which they operate (Wells and Falcone, 2008). They also experience even greater resource constraints than their nontribal counterparts do. Although workloads of tribal agencies are increasing, available staffing, facilities, and equipment remain inadequate (Wakeling et al., 2001), leaving many agencies unable to provide adequate coverage of police services around the clock and across the wide geographic area of the reservations.

Multiple layers of legal authority (federal, state, and tribal) create additional complexities that confuse jurisdiction over criminal matters, reducing overall accountability. Public Law 83-280, 1953, which mandated transferring legal authority from the federal to state governments in six states (California, Minnesota, Nebraska, Oregon, and Wisconsin and, when it became a state, Alaska) and is optional in others, has led to receding federal involvement in tribal criminal justice affairs and the increasing intervention of state criminal justice agencies. Tribes in mandatory Pub. L. 83-280 states are less likely to have their own law enforcement agencies than those in optional– and non–Pub. L. 83-280 states, and tribes in Pub. L. 83-280 states are less likely to be funded through Bureau of Indian Affairs (BIA) contracts.

Courts

Like with tribal law enforcement, tribal courts face unique challenges regarding funding and jurisdictional authority, as well as other issues that small and rural court systems experience. Native American tribal courts (which tribes operate) and Courts of Indian Offenses (which the BIA operates) resolve disputes, including civil and criminal matters, in Native American country (Jones, 2000). Limits on tribal court authority vary depending on the types of people and cases involved. Jurisdictional disputes between federal, state, and tribal authorities have been improved by a variety of strategies, including state–tribal court forums, joint-jurisdiction courts, and written collaborative agreements between state and tribal authorities (Arnold, Reckess, and Wolf, 2012).

In addition, culturally specific programs have emerged in state courts; these include sentencing circles based on restorative justice principles (see also Mirsky, 2004a; Mirsky, 2004b) elder mentoring for court-involved youth (Arnold, Reckess, and Wolf, 2012), and some wellness courts for substance use (Tribal Law and Policy Institute, 2003). Child welfare practices that bridge state and tribal courts include subsidized guardianship by tribal kin in lieu of adoption, wraparound family services in abuse and neglect cases, and compliance with the Indian Child Welfare Act (Pub. L. 95-608, 1978), which seeks to prevent removal of Native American children from their families. Finally, professional training and outreach with bar associations can help bridge state and tribal divides by promoting knowledge about culture, law, and best practices with tribal courts (Arnold, Reckess, and Wolf, 2012).

Corrections

Research on tribal corrections agencies has generally focused on institutional corrections because data and published literature on existing tribal probation agencies are rare (Cobb, 2011; Melton, Lucero, and Melton, 2010). There is a critical need for effective transitional and reentry programs in tribal areas given higher rates of violence, substance use, victimization, and poverty. According to *Census of Tribal Justice Agencies in Indian Country, 2002* (Perry, 2005),[2] 69 percent of tribes with judicial systems (*n* = 188) and 40 percent of all tribes participating in the census (*n* = 314) utilized probation for adults (*n* = 130) or juveniles (*n* = 124). Formal tribal probation is a relatively recent concept; tribes have historically dealt with law breaking by utilizing a so-called peacemaking approach, in which a respected member of the tribe, such as a tribal elder, assists in mediating an outcome that compensates the victim for harm done (Cobb, 2011). One challenge is to integrate these kinds of culturally competent approaches with best practices in probation while formal tribal justice systems continue to evolve. Chronic underfunding and coordination with federal, state, and local agencies present additional challenges. Researchers and practitioners have increasingly attempted to bridge the gap by providing guidance on program development (Melton, Lucero, and Melton, 2010) and best practices for personnel, such as judges (Cobb and Mullins, 2009; Cobb and Mullins, 2010) and probation officers (Cobb, 2011), as well as raising awareness of risk–need responsivity issues for Native Americans involved in the criminal justice system (Cobb, 2013; Melton, Cobb, et al., 2014).

A 2004 study by the U.S. Department of Justice (DOJ) Office of the Inspector General (OIG) found that the BIA "failed to provide safe and secure detention facilities throughout Indian Country" (U.S. Department of Interior, 2004, p. 3). The majority of serious incidents, including fatalities, attempted suicides, and escapes, not only threatened safety and security but also were almost entirely unknown to the BIA. In addition, staffing levels, facility maintenance, management and accountability of funds, and training of personnel were deemed inadequate, failing to meet minimum operational standards. Although funding for detention increased 48 percent in the five years following the 2004 OIG report, a 2011 evaluation found that staffing, facility maintenance, and financial management and accountability remained inadequate (U.S. Department of Interior, 2011).

According to the BJS annual survey Jails in Indian Country, the number of tribal jails operating within federally designated Native American country increased from 68 to 79 between 2004 and 2014, and the number of inmates held in tribal jails increased from 1,745 to 2,380 (Minton, 2015). There were more than 10,000 admissions to tribal jails between midyear 2013 and mid-year 2014—a slight decline from the previous years. Growth in facilities led to a decline in the overall percentage occupied, from 81 percent to 64 percent. More than 75 percent of all tribal detention facilities are small, having fewer than 50 beds; these facilities tend to have higher occupancy rates than their larger counterparts (Minton, 2015).

Such efforts as the Tribal Law and Order Act of 2010 (Pub. L. 111-211) are intended to enhance tribal justice systems by establishing culturally specific programs and alternatives to incarceration, as well as facilitating intergovernmental coordination and collaboration.

---

[2] Unfortunately, these data are more than ten years old and likely to be considerably outdated. For example, at the time of the 2002 census, there were only 341 federally recognized tribes in the United States. As of 2013, this number was 566.

Border Agencies

A less coherent body of research has explored the unique needs of law enforcement, courts, and corrections agencies in border areas. Unlike areas with small and rural agencies and many tribal agencies, border areas are both rural and urban, with agencies covering jurisdictions in a range of sizes. The above discussion on the challenges that small and rural agencies face is thus also relevant for many border agencies. The common thread among these agencies, however, is the need to address specific border-related issues, such as homeland security, human trafficking and undocumented immigration, and the trafficking of illicit drugs and cash flows across the border (National Law Enforcement and Corrections Technology Center [NLETC], 2011). Increased collaboration and information-sharing with federal authorities might also be required because of the nature of the issues with which local criminal justice agencies deal in these areas.

Technology in the SRTB Context

A substantial literature exists on the use of technology in criminal justice agencies, but most of that work focuses on law enforcement agencies in larger and more-urban settings (e.g., Gordon et al., 2012; Homeland Security Studies and Analysis Institute, 2012; International Association of Chiefs of Police [IACP], 2005; Koper, Taylor, and Kubu, 2009; LaTourrette et al., 2003; Law Enforcement and Corrections Technology Advisory Council [LECTAC], 2009; NIJ, 2009; Police Executive Research Forum, 2011; Schwabe, Davis, and Jackson, 2001). Similar limitations are found in the existing literature on the use of technology in corrections (e.g., Atherton and Russo, 2009; LECTAC, 2009; LIS, 1995; NIJ, 2009; Scism, 2009; Zelenak and Goff, 2005) and the courts (e.g., Spangenberg et al., 1999; Hill, 2004; Lambert, 2008; Gallas, 2001; Courts Task Force, 2001; NIJ, 2009).

These previous studies have addressed applications as varied as the following:

- information-collection tools and sensors, ranging from routine (e.g., closed-circuit television in prisons) to specialized (e.g., wiretapping or tracking devices)
- information-management tools, such as record-management systems (RMSs) used for everything from evidence to court case files
- analytic tools to leverage available data in investigation and decisionmaking
- communication tools used within facilities, with specific populations, and with the public
- manned vehicles and remotely piloted ground, air, and water systems
- weapon, tactical, and force innovations for intervention or site security
- training and protective technologies, from stab-resistant vests to online teaching tools.

There is also a limited literature measuring the *effects* of these technologies in the criminal justice fields, including the value they provide through increased effectiveness, the more-efficient use of resources, and other outcomes, such as officer and employee safety (for overviews, see Bean, 1999, and Byrne and Marx, 2011). For example, research indicates that IT upgrades in police departments are associated with improvements in perceptions of productivity, efficiency, and capability (Brown, 2001). Other research has shown that police departments that implement IT have fewer officers per capita (Nunn, 2001), which might indicate greater efficiency. On the other hand, IT might be associated with improved productivity only when coupled with complementary organizational and management practices (Garicano and Heaton, 2010).

A better understanding of how technologies are used, the outputs and outcomes of their use, and how they could be used to address unmet needs is critical for prioritizing the acquisition of the many possible criminal justice technologies and applications. However, it is equally important to understand the unique needs of the SRTB justice systems, which are often very different from those in larger or more urban environments.

Selecting clear and defensible technology priorities has been a challenge in the field of technology planning overall, not just within criminal justice communities. These difficulties have arisen from the uncertainty associated with predicting technology developments and the ways in which their adoption will affect organizations or societies, as well as from the challenge of providing a common basis for comparing technology's effects on different users and sectors. Examples of past technology prioritization and planning efforts for the criminal justice and broader emergency responder communities include Garwin, Pollard, and Tuohy, 2004; Gordon et al., 2012; IACP, 2005; Koper, Taylor, and Kubu, 2009; LaTourrette et al., 2003; LECTAC, 2009; Courts Task Force, 2001; NIJ, 2009; and Royal, Donahue, and Kirby, 2008.

Conclusion to Discussion of the SRTB Context

This discussion of the SRTB context for criminal justice agencies has highlighted the limited amount of research that has been done in these areas. JIC's purpose is to gather additional information on the challenges that SRTB agencies face, identify relevant technology solutions that can address those challenges, and assess these technology solutions as they are implemented in real-world situations. These activities will provide actionable guidance to SRTB agencies for prioritizing, planning, and implementing technology.

The Justice Innovation Center's Four Research Goals

JIC's main activities have been planned around four research goals, which contribute to the overall mission of the center. We planned tasks to support these goals using a flexible framework that allows work begun in year 1 to continue into the following years. The four goals guiding JIC's work are as follows:

1. Identify the unique operational needs of SRTB justice systems.
2. Identify potential, innovative, technology-based solutions to meet those needs.
3. Conduct trials to evaluate those solutions' impact on criminal justice policy and practice in representative agencies.
4. Effectively disseminate the results of those evaluations to inform policy and practice in SRTB justice systems across the United States.

Some tasks associated with each of these research goals have occurred in parallel and will be repeated in appropriate combinations and applied to topics specific to the agency types (law enforcement, the courts, and corrections) or sectors (small, rural, tribal, and border), as well as topics and technologies that cut across these communities and settings. JIC's efforts thus far have concentrated on research goals 1 and 2, and the findings from those efforts provide the basis for the current report.

Research Goal 1: Identify Small, Rural, Tribal, and Border Needs

To identify the unique operational needs of the SRTB justice systems—research goal 1—JIC research staff undertook three main tasks: literature review, attendance at conferences, and interviews with agency representatives. These three sources of information allowed JIC to identify and integrate available data on known or emerging needs or operational problems in SRTB agencies. These include unmet needs in the communities of interest that (1) lack solutions, (2) have existing solutions that are not performing to expectations because of the solutions' limitations, and (3) have existing solutions that are not effective because of other barriers. Each of these components helps to define the demand side of the market for technology in these communities.

First, we conducted an in-depth literature review specifically focused on technology use in the three main agency types (law enforcement, courts, and corrections) and in the four main sector types (small, rural, tribal, and border), which define our 12 different communities of practice (see Table 1.1 for definitions). This information helped orient members of the JIC advisory panel in identifying and prioritizing their technology needs and will continue to inform scans of existing technologies. Some areas of the literature proved to be quite rich, while others were much less so; Chapter Three presents a detailed discussion of the results of the literature review and their implications for JIC work.

Second, JIC's research team attended a variety of conferences to introduce JIC to a range of audiences. JIC staff held individual meetings with key stakeholders at each conference and participated in panel presentations on JIC's planned work. The conference attendance also allowed JIC researchers to better understand the audiences for planned research efforts, make valuable connections with practitioners who might be potential participants in future JIC efforts, and learn about technology being promoted to various types of agencies through vendor exhibits.

Third, we conducted in-depth semistructured interviews with agencies in all three agency types and all four sector types. We organized the interviews across several domains to help identify SRTB agencies' greatest unmet operational needs and solutions already in use. Domains covered in the interviews included operational needs, technology solutions, and dissemination and TA needs. Between June and October 2015, we conducted a total of 147 interviews. We then coded the interviews and analyzed them to identify key needs of stakeholders in the agencies and sectors of interest to JIC. The findings from the interviews helped JIC to solidify its understanding of current needs among relevant agencies, informed the JIC advisory panel discussions on needs and potential technology solutions, and helped JIC researchers identify potential sites and participants for future technology pilots. Chapter Two describes the meth-

Table 1.1
The Justice Innovation Center's 12 Communities of Practice

Agency Type	Sector Type			
	Small	Rural	Tribal	Border
Law enforcement	1	2	3	4
Courts	5	6	7	8
Corrections (institutional and community)	9	10	11	12

ods used for identifying and recruiting interviewees; Chapter Four provides the results from the interviews.

Research Goal 2: Identify Technologies

JIC's initial focus under this research goal was on extant technologies already being implemented in different contexts (e.g., large or urban settings) that might meet the needs of the SRTB agencies. We expect that, in subsequent years, JIC's work will focus on adapting the technology to the specific needs of the SRTB settings, adapting technologies from other fields that address the needs of criminal justice agencies in the SRTB settings, and exploring completely new technologies to address SRTB needs where none already exists.

To identify potentially innovative technology-based solutions to meet the needs of SRTB agencies, we conducted three main tasks. First, we coordinated with other NIJ grantees to leverage work already conducted in this area with relevance to JIC's goals, including work conducted by an NIJ-funded center at RAND. This ongoing project was tasked with identifying the highest-priority criminal justice technology needs for law enforcement, corrections, and courts. A multiorganization partnership is conducting that work, which includes both quantitative and qualitative measures of the technological, administrative, and community barriers to effective technology adoption. Methods designed at RAND for that project's technology prioritization work proved invaluable to JIC's work, providing the groundwork to effectively carry out JIC's research aims. That project also informed analysis of the interview findings and provided a larger context within which to understand the needs of SRTB agencies.

Second, we built on the center's findings on high-priority criminal justice technology needs, the literature review, and the interview findings to develop a comprehensive list of technologies used in the criminal justice system. The compilation of technologies can be seen as a living list that will be updated as the work of JIC proceeds and as new technologies emerge as relevant to SRTB criminal justice agencies. JIC's understanding of existing technologies informed analysis of the interview findings, development of materials for and the discussions during the advisory panel meeting, and plans for potential technology pilots.

For the third task, RAND proposed to conduct an expert panel with practitioners, technical experts, and stakeholders to identify and rank technology needs. We held these discussions during an advisory meeting in December 2015, for which more than 40 people, including practitioners and researchers representing all agency and sector types, convened in Arlington, Virginia. Participants, divided by agency type, first participated in extensive discussions that JIC researchers facilitated on the challenges or issues that agencies like theirs face and potential technology-based solutions or strategies to address those issues. Next, participants ranked the potential solutions in order of importance. Chapter Five provides details on the advisory panel.

Research Goal 3: Implement Technologies

The next step for JIC will be to select technologies that meet an identified need and have the potential, if effective, to have a significant impact on SRTB agencies. Selected technologies will be implemented in pilot sites and evaluated in order to provide guidance to other agencies as to the utility of and costs associated with the technology. Because of delays at the start of year 1, JIC could not begin implementation of any technology pilots during year 1. However, the work done in JIC's first year provided insight into potential technologies to be piloted and, through such activities as conference attendance, interviews, and the advisory panel, served to connect

JIC with practitioners who are interested in participating in such pilots. Planning for technology pilots began at the end of year 1 and implementation began in year 2.

Research Goal 4: Disseminate Findings

The fourth goal of JIC is to disseminate its findings in order to ensure its utility to a wide audience of practitioners, researchers, and government agencies, especially those in SRTB contexts. Effective dissemination and integration with the communities of interest are essential to inform NIJ funding strategies and priorities for SRTB criminal justice system technology research and development; to facilitate the widespread education, adoption, and routinization of best practices among justice practitioners and technology providers; and to stimulate research among organizations whose work is positioned to further support SRTB agencies' ongoing needs. We believe that a research base and evidence of the effectiveness of technology is necessary before findings, recommendations, or TA is provided to the field.

JIC worked with another NIJ contractor to redesign and update the content of a JIC website (Justice Technology Information Center, undated) (https://www.justnet.org/about/jic-center.html) (the website was based on one already in use for JIC's predecessor). The website is designed with a combined top-down/bottom-up approach. The top-down approach encompasses validated information on resources and best practices provided by a centralized clearinghouse, which will be particularly relevant once results from the technology pilots are finalized. The bottom-up approach focuses on facilitating communication between local agencies and encouraging local stakeholders to share ongoing questions and local solutions within and between agencies. Such an approach was evident in JIC's advisory panel meeting that brought together practitioners from all agency and sector types to discuss their most-salient needs in person. Such meetings inform, and provide the basis for, additional online communication between agencies.

The same contractor that is responsible for the website redesign also worked with JIC to create a new logo in order to provide cohesive branding for the website and any printed materials that might be distributed. To engage with interested audiences online, JIC also registered a Twitter account, which it will use once JIC products are released. As dissemination of findings begins in earnest, the Twitter account will be used regularly, and the website content will be updated frequently.

In future years, as JIC's work matures, additional dissemination vehicles will be used, including possible newsletters, briefs, and searchable databases of technology and relevant literature. The pace of dissemination is expected to increase as work under JIC continues, and dissemination will be a key center activity throughout its existence.

Organization of This Report

The remainder of this report consists of five chapters:

- Chapter Two describes our methodology.
- Chapter Three provides the results of our literature review.
- Chapter Four describes the results of our interviews with agency personnel.
- Chapter Five discusses the results of the JIC advisory panel.
- Chapter Six provides our conclusions.

There are also five appendixes:

- Appendix A reproduces the letters of invitation and support and the JIC mission statement.
- Appendix B provides the interview coding framework.
- Appendix C lists members of the 2015 JIC advisory panel.
- Appendix D provides the advisory panel read-ahead material.
- Appendix E lists tier 1 needs from the JIC advisory panel, by agency type.

CHAPTER TWO
Methodology

In this chapter, we describe the methodology used in this research. We begin with our definitions of SRTB agencies. Then we detail our methods for conducting the literature review. Next, we discuss the semistructured interviews we used in the study to document the challenges that SRTB criminal justice agencies face. Finally, we describe the methods we used to rank and prioritize unmet technology needs in the communities of practice.

Defining Communities of Practice: SRTB Criminal Justice Agencies

In this section, we describe how we identified and defined SRTB criminal justice agencies. We then describe, based on available empirical literature, how these agencies differ from larger urban agencies.

Our focus is on law enforcement agencies and sheriff's departments, district and county courts (of limited or general jurisdiction), local jails, and probation agencies. We excluded all federal agencies, state police, state prisons, and most parole agencies because they are typically much larger and better resourced than agencies operating at a county or municipal level. Our definitions of rural, border, and tribal agencies apply across all agency types, while the definition of small agencies is dependent on agency type. Table 2.1 summarizes the definitions of the communities of practice used in this report.

Small Agencies
Law Enforcement
For the purposes of this project, we define a *small* law enforcement agency as one with fewer than 25 full-time sworn law enforcement officers. This definition is similar to previously published definitions, which have defined agency size based on either the number of personnel employed or the residential population served. In this study, we focused on full-time sworn officers because there is a great deal of variation in the use of civilian personnel across departments (Crank, 1989). For example, a previous study defined a small (or rural) law enforcement agency as one with fewer than 20 sworn officers serving an area with fewer than 50,000 residents (NIJ, 2004), whereas another defined a small law enforcement agency as one with between one and 25 full-time officers (Burruss et al., 2014). Similarly, the Commission on Accreditation for Law Enforcement Agencies defines a small agency as one with 24 or fewer personnel, and the IACP's Smaller Law Enforcement Agency Program includes departments serving fewer than 50,000 residents.

Table 2.1
Definitions for Communities of Practice

Agency Type	Small	Rural	Tribal	Border
Law enforcement	Law enforcement agency with fewer than 25 full-time sworn officers	Law enforcement agency located in a county with fewer than 2,500 residents and nonadjacent to a metro area	Law enforcement agency operated by a tribal entity	Law enforcement agency located in a border county
Court	Court serving a district with fewer than 100,000 residents	Court located in a county with fewer than 2,500 residents and nonadjacent to a metropolitan area	Court operated by a tribal entity	Court located in a border county
Institutional corrections	Jail with an average daily population of less than 50	Jail located in a county with fewer than 2,500 residents and nonadjacent to a metro area	Jail or other detention facility operated by a tribal entity	Jail or other state- or locally run detention facility located in a border county
Community corrections	Probation or parole agency serving districts with fewer than 25 full-time officers	Probation or parole agency located in a county with fewer than 2,500 residents and nonadjacent to a metro area	Probation or parole agency operated by a tribal entity	Probation or parole agency located in a border county

Of the 17,985 agencies in the CSLLEA, 73 percent (13,096) have fewer than 25 sworn officer positions and fit our definition of a small agency. In the current sample of agencies participating in interviews, the small agencies averaged 16 sworn-officer positions (although all positions might not be filled), ranging from a capacity of three to 26 officers.

Courts

We define a *small* court as one serving a district with 100,000 or fewer residents. Lack of available court statistics precluded us from defining a small court by staffing or personnel, caseload, or some other metric.[1] We searched the available literature for definitions of small courts and found only one, which was used in an early BJS report on state court prosecutors in small districts (DeFrances, 2003). That definition included all state-level court districts with 250,000 or fewer residents, which we determined would include urban areas and some cities that were beyond the scope of JIC. We thus used a smaller population to identify small courts for JIC.

Institutional Corrections

We define a *small* jail as one with an average daily population of fewer than 50, which is consistent with the American Jail Association (AJA) definition of a small jail as one with fewer than 50 beds (AJA, undated). *Census of Jail Facilities, 2006*, enumerated 2,949 local jails in the United States; of these, nearly 18 percent (*n* = 528) fit this definition of *small* (Stephan and Walsh, 2011).

---

[1] In practice, the small-only courts used in this research served areas that ranged in size from 15,000 to about 40,000 residents. Some rural courts were even smaller, serving counties with as few as 2,000 residents.

Community Corrections

There are few available data on community corrections agencies. We know of no national directory of community corrections agencies that could be used to develop a population from which to draw an interview sample; the Bureau of Justice Assistance (BJA) conducts annual probation and annual parole surveys, but those are focused on offender counts and characteristics, not on agencies. We thus identified agencies meeting our SRTB definitions by searching for contact information on state directories, where those existed, or by searching for individual agencies in counties meeting the rural and border criteria. We also learned through our interviews (discussed in Chapter Four) that small agencies are typically consolidated into multicounty districts. We therefore focused on rural, tribal, and border probation departments to define our community corrections sample. Many consolidated districts are likely made up at least in part of counties that might be considered small in terms of the probation population served—e.g., a small county pools resources with a neighboring larger county or a network of small counties as part of a larger group—so we chose not to filter candidates based on a potentially misidentified definition of small community corrections agencies for this research.

Where small-population counties share district resources with larger-population counties, our interviews focused on topics that were unique or germane to SRTB county technology needs.[2] In cases in which all district member counties are rural, tribal, or border, interview topics included challenges related to coordination across counties. By not defining small agency, we avoid erroneously excluding any agency that might serve a geographically large area made up of counties that would serve small numbers of clients absent that consolidation. Further, we are unlikely to miss applicable agencies given the significant degree of overlap between rural and small agencies—most small agencies are in rural areas across all justice domains. However, it is possible, though probably unlikely, that our sample construction obfuscates technology needs that are unique to small agencies alone. The fact that small agencies tend to consolidate reduces the risk of this limitation.

Table 2.2 presents an overview of how we defined small and tribal agencies—the sectors for which the definitions varied across agency types—within each segment for the purposes of

Table 2.2
Definitions of Small and Tribal Agencies

Agency Type	Small	Tribal
Law enforcement	<25 sworn officers (CSLLEA)	Independent agency run by federally recognized tribe (CSLLEA)
Court	Population <100,000 (census)	Independent agency run by federally recognized tribe (NAICJA)
Institutional corrections	Jail population <50 (Census of Jail Facilities)	Operated by tribal authority *or* the BIA (Jails in Indian Country survey)
Community corrections	Not applicable; grouped into multicounty districts	Independent agency run by federally recognized tribe (CSLLEA)

NOTE: NAICJA = National American Indian Court Judges Association.

---

[2] Small population—in part the basis for our definition of rural—is a seemingly appropriate proxy for small probation programs given the lack of data on probation department characteristics.

this project. In the rest of this section, we discuss how we developed those definitions and the source of the metrics used.

Rural Agencies

There are multiple ways to define *rural* in the literature—the Rural Policy Research Institute identified more than 15 definitions of *rural* in use by federal programs (Coburn et al., 2007). In general, this term refers to areas with low-density populations at a specified distance from large cities and metropolitan areas. We define a *rural* law enforcement agency as one located in a county characterized as having (1) a census-defined rural population, (2) a census-defined urban population under 2,500, or (3) a census-defined urban population of between 2,500 and 19,999 in an area not adjacent to a metropolitan area. This definition was developed in part based on USDA's nine-point rural–urban continuum codes (Economic Research Service, 2013). According to the 2008 CSLLEA, 17 percent of law enforcement agencies (3,020) were located in rural areas. According to *Census of Jail Facilities, 2006*, more than 27 percent of the 2,949 local jails enumerated were rural (n = 795). However, there is no data source that provides a count of courts or community corrections agencies located in rural areas.

Tribal Agencies

There are 566 federally recognized tribes in the United States (BIA, 2013). For this study, we define a *tribal* agency as a local law enforcement agency that one or a confederation of federally recognized tribes operates independently. Tribal agencies also include those, such as some tribal jails, that the BIA operates.

In the 2008 CSLLEA, about 1 percent of all law enforcement agencies (178) were considered tribal agencies. According to the National Directory of Tribal Justice Systems, which NAICJA manages, there are 259 active and 328 nonactive or in-development tribal courts (total = 587). This is somewhat larger than previous estimates; for example, of the 314 tribes that responded to the 2002 Census of Tribal Justice Agencies, 188 reported operating some form of judicial system, and 175 reported operating tribal courts (Perry, 2005). The most-recent Jails in Indian Country report enumerated 79 jails operating in Native American country by tribal authorities or the BIA (Minton, 2015). This is a slight increase from the 2002 Census of Tribal Justice Agencies, which reported that 71 tribes reported operating jails (Perry, 2005). Finally, 130 tribes reported operating probation for adults, and 124 reported operating probation for juveniles (Perry, 2005).

Border Agencies

We define a border agency based on proximity to either the Canadian or Mexican border using available census data. There are 368 counties located within 100 miles of the U.S. border with Canada or Mexico—109 of which are located in one of 17 states directly adjacent to the border. About 14 percent of law enforcement agencies in the 2008 CSLLEA (2,532) were located in border counties, as were nearly 12 percent of jails (350) in *Census of Jail Facilities, 2006*. Like with rural areas, there is no data source that provides a count of courts or community corrections agencies located in border areas.

Literature Review Approach

To identify literature addressing technology needs and use in SRTB agencies, we conducted a search under the three main criminal justice agency types: law enforcement, courts, and corrections. To identify literature, we undertook two main efforts: A RAND librarian conducted a search for relevant literature in several available databases, and researchers on the JIC team conducted searches using Google Scholar. We included literature in peer-reviewed journals, as well as items considered to be part of the gray literature—such as articles from trade publications, unpublished papers, technical reports or technical documentation, or theses.

The RAND librarian searched the following databases:

- Academic Search Complete contains thousands of journals in sciences, arts and humanities, and social sciences.
- Business Source Complete contains business and economics journal articles, country profiles, and industry reports.
- Criminal Justice Abstracts contains abstracts related to criminal justice and criminology, including current books, book chapters, journal articles, government reports, and dissertations published worldwide.
- National Criminal Justice Reference Service's database contains summaries of the more than 185,000 criminal justice publications housed in the NCJRS Library collection.
- Scopus contains peer-reviewed research and web sources in scientific, technical, medical and social sciences, and arts and humanities areas.

We used the following search terms in both the database search and the Google Scholar search:

> ((("law enforcement" OR police OR policing OR crime OR "criminal justice" OR justice OR court OR courts OR jail OR prison OR prisons OR corrections)) AND KEY ((rural OR nonurban OR tribal OR border OR SRTB OR "smaller community" OR "smaller communities" OR "small community" OR "small communities" OR "small agency" OR "small agencies" OR "smaller agencies" OR "small agency"))

This database search produced more than 350 citations. The Google Scholar searches produced a significantly larger number of sources, but many of those beyond the initial 50 to 100 sources had low levels of search relevance (i.e., were off-topic), so we excluded them from the review. We reviewed the abstracts of the initial set of citations identified and selected approximately 100 sources that were relevant to the JIC literature review—covering technology and relevance to SRTB agencies. We also included some that did not explicitly discuss SRTB agencies or contexts but that addressed issues of relevance to those agencies or to criminal justice agencies in general. We excluded any source that did not discuss the topics of interest, did not address technology-related topics, or was an opinion piece. Also, the search prioritized more-recent literature over literature that had been published prior to 2000, unless we identified the older piece as a seminal or key article on a specific topic.

We have included in a reference database the sources located thus far and will update it if we identify additional literature later. The reference list for this report includes the references cited.

Semistructured Interviews

Research into the needs of SRTB agencies and effective solutions to meet those needs is sorely lacking. To fill this gap, JIC researchers interviewed a national sample of SRTB practitioners in all of the main agency types of interest about the unique operational challenges that these agencies face. We also interviewed selected justice technology experts to get their perspectives on how to best address some of these unique challenges.

After determining the universe of SRTB agencies, we developed and tested interview protocols. Once the protocols were final, we sent invitations to randomly sorted agencies in the sample frames and asked them to participate in the study. Because there is no one sampling frame, there was no way for us to assess the overall representativeness of our final sample. As we completed the interviews, we transcribed and coded the responses. We then analyzed these coded excerpts for trends both across and within agency types. We then presented the results to an advisory panel for feedback and guidance on determining which of the identified needs should be JIC's focus in year 2. In this section, we describe the methods used to develop the interview protocol, assess our target agencies, and collect and analyze the data.

Interview Protocol

We organized the semistructured interview protocol that we developed into three domains: (1) key operational challenges; (2) potential technological and operational solutions to those challenges; and (3) information on technology education, acquisition process and procedures, and dissemination needs. For each of the three sections of the interview, we solicited open-ended responses first. After the respondent appeared to have provided a complete response, the interviewer used concrete examples to probe the respondent's memory for other potential challenges or solutions that might be applicable in the person's field. Guided by this basic interview structure, interviewers were free to deviate from the protocol as necessary to further enrich the discussion.

Interview Process

JIC researchers who trained on the interview protocol conducted 147 interviews across the three agency types—law enforcement, courts, and institutional and community corrections in SRTB areas. To identify a sample of representative agencies, we used existing databases from a variety of sources: professional associations and their mailing lists and a snowball sample that started with our own contacts among the target agency types. The method of identifying interviewees varied slightly for each agency type, depending on what kinds of information were available for relevant agencies. Later in this section, we describe the methods for sample development for each of the 12 specific communities of practice.

We sent letters via email inviting police chiefs, sheriffs, court administrators, and corrections officials to participate in the interviews, along with letters of support from NIJ and the JIC mission statement. We asked respondents to contact JIC staff via project email addresses unique to each agency type to schedule interviews. Once an interview was scheduled, we sent that respondent a confirmation email. Appendix A reproduces the invitation materials.

We sent invitations on a rolling basis until we had completed the target number of interviews (eight) for each agency and sector type. In all but one case, we were able to complete the target number of interviews; for tribal community corrections, we could conduct only one interview because of difficulties identifying and establishing contact with these agencies. After

Table 2.3
Number of Interviews Conducted, by Agency and Sector Type

Agency or Sector Type	Law Enforcement	Court	Institutional Corrections	Community Corrections	Total
By agency type, total					
All small agencies	23	10	13	0	46
All rural agencies	8	11	27	31	77
All tribal agencies	9	9	8	1	27
All border agencies	12	8	20	10	50
By sector type, broken into combinations					
Small only	9	8	1	—	18
Rural only	2	7	11	26	46
Tribal only	4	9	1	—	14
Border only	5	5	8	5	23
Small and rural	4	1	5	—	10
Small and tribal	4	—	—	—	4
Small and border	4	—	3	—	7
Small and rural and tribal	—	—	2	—	2
Small and rural and border	2	1	—	—	3
Rural and tribal	—	—	1	—	1
Rural and border	—	2	5	4	11
Rural and tribal and border	—	—	1	1	2
Tribal and border	1	—	1	—	2
Small and rural and tribal and border	—	—	2	—	1
Expert interviews	1	2	—	—	3
Total interviews	36	35	41	36	148

NOTE: Agency types are not mutually exclusive because a few agencies fit into multiple categories. No agencies were small, tribal, and border, so we exclude this category. — = We did not conduct an interview in this category.

completing the target number of interviews, we sometimes conducted interviews with additional agency personnel who sought to participate, resulting in additional interviews for several agency–sector categories.

Tables 2.3 through 2.6 show the breakdown of agency and sector type, along with geographic distribution. Table 2.3 provides the number of interviews conducted, by agency and community of practice type. Many agencies fall into more than one category. Table 2.4 shows interviews by state and agency type.

Tables 2.5 and 2.6 show the total number of agencies of each community-of-practice type in the sampling frame and the percentage interviewed.

Table 2.4
Interviews Conducted, by State and Sector Type

State	Small	Rural	Tribal	Border	Total Interviews Conducted
Alabama	—	—	—	—	—
Alaska	2	3	2	3	7
Arizona	2	1	7	5	12
Arkansas	—	1	—	—	1
California	1	4	1	5	10
Colorado	1	4	1	—	4
Connecticut	—	—	—	—	—
Delaware	—	—	—	—	—
Florida	1	—	—	—	1
Georgia	1	—	—	—	1
Hawaii	—	—	—	—	—
Idaho	—	—	2	—	2
Illinois	—	2	—	—	2
Indiana	1	1	—	—	2
Iowa	2	3	—	—	3
Kansas	—	2	—	—	2
Kentucky	—	1	—	—	1
Louisiana	1	—	1	—	1
Maine	—	—	—	2	2
Maryland	—	—	—	—	—
Massachusetts	—	—	—	—	—
Michigan	—	2	—	2	3
Minnesota	6	8	2	6	13
Mississippi	—	—	—	—	—
Missouri	1	—	—	—	1
Montana	1	4	1	3	4
Nebraska	—	6	1	—	7
Nevada	1	1	—	—	2
New Hampshire	1	1	—	2	3
New Jersey	1	—	—	—	1
New Mexico	2	2	2	2	4
New York	2	—	—	2	4

Table 2.4—Continued

State	Small	Rural	Tribal	Border	Total Interviews Conducted
North Carolina	1	2	—	—	2
North Dakota	3	5	1	4	6
Ohio	1	2	—	2	3
Oklahoma	1	1	1	—	2
Oregon	—	—	—	—	—
Pennsylvania	—	1	—	1	2
Rhode Island	—	—	—	—	—
South Carolina	1	2	—	—	3
South Dakota	1	2	1	1	2
Tennessee	—	—	—	—	—
Texas	3	1	—	2	4
Utah	—	—	—	—	—
Vermont	—	—	—	—	—
Virginia	1	11	—	—	11
Washington	3	1	2	8	10
West Virginia	1	—	—	—	1
Wisconsin	2	3	2	—	5
Wyoming	1	—	—	—	1
Total	46	77	27	50	145

NOTE: One tribal agency spanned both Arizona and California; we include it here as California. — = We did not conduct an interview in this category. Rows do not sum to the number in "Total Interviews Conducted" because an agency can belong to multiple categories. We contacted agencies in all 50 states; the distribution in this table reflects those that responded to our invitation to participate.

We audio-recorded, transcribed, and coded the interviews using the Dedoose qualitative software platform (SocioCultural Research Consultants, 2015). A six-member coding team developed a first-level coding framework, with each member trained in the interview protocol for at least one of the agency types. We designed the codes to apply to all agency types, with superordinate, basic, and subordinate levels. The framework development drew on a previous RAND study on priority criminal justice needs (Hollywood, Boon, et al., 2015; Silberglitt et al., 2015), which developed a law enforcement–related lexicon. The codes generally captured the past, present, and future operational needs and challenges that agencies face, and we found the law enforcement codes that were previously developed largely worked across agency types for the SRTB interviews. Appendix B provides the interview coding framework.

To validate the utility of the coding framework, team members independently test-coded one interview from each agency type and added or adjusted codes based on information particular to the SRTB interviews. When a common understanding of the coding framework was

Table 2.5
Agencies, by Community of Practice: Law Enforcement and Courts

Sector	Law Enforcement Number	Law Enforcement Percentage Interviewed	Courts Number	Courts Percentage Interviewed
Small	13,096	0.2	—	—
Rural	3,020	0.3	—	—
Tribal	178	5.6	259	3.5
Border	2,532	0.4	—	—

SOURCES: Number of law enforcement agencies by sector: Reaves, 2011. Number of tribal courts: NAICJA, undated. Number of tribal community corrections services: Perry, 2005.

NOTE: — = not applicable because no comprehensive count of agencies was available.

Table 2.6
Agencies, by Community of Practice: Institutional and Community Corrections

Sector	Institutional Number	Institutional Percentage Interviewed	Community Number	Community Percentage Interviewed
Small	528	2.5	—	—
Rural	795	3.4	—	—
Tribal	79	10.1	130	0.8
Border	350	5.7	—	—

SOURCES: Number of law enforcement agencies by sector: Reaves, 2011. Number of institutional corrections facilities by sector: Stephan and Walsh, 2011. Number of tribal courts: NAICJA, undated. Number of tribal community corrections services: Perry, 2005.

NOTE: — = not applicable because no comprehensive count of agencies was available.

established between team members, interviews were fully coded; if information was mentioned in the text that a code did not capture, a new appropriate code was created. Throughout the coding period, coders held weekly progress meetings to discuss any changes or issues. These meetings served to address uncertainties as to which code was the most appropriate for the excerpt in question and, if information was mentioned in the text that a code did not capture, whether a new appropriate code should be created.

Once the interviews had been coded, data were also used to document common and unique operational needs across the different systems and settings, reported in Chapter Four. We compiled these themes for review and used them to develop content for the advisory panel; they will also aid JIC in selecting high-priority technology evaluation projects likely to have the greatest impact on SRTB agencies. The next sections describe sample identification for each community of practice, organized by agency type.

Sample Development
Law Enforcement Sample
We used a three-pronged approach to identify our sampling frame and aid in the recruitment of representative subjects. This approach included the use of (1) established databases from BJA, including the CSLLEA; (2) professional associations and their corresponding mailing lists, such as the National Sheriffs' Association; and (3) a carefully constructed snowball sample starting with our own contacts in the criminal justice system. The contacts arose mainly through other projects in SRTB locations in California and Arizona. The law enforcement sample selected for inclusion corresponded to the defined SRTB criteria outlined earlier in this chapter.

The defined SRTB categories are not mutually exclusive; as shown in Table 2.4, there is considerable overlap among the SRTB law enforcement agencies in our sample. As of the last available CSLLEA (2008 [Reaves, 2011]),[3] more than one in five small agencies was located in a rural county, and nearly one in seven was located in a border area. Conversely, more than 95 percent of rural agencies, 85 percent of tribal agencies, and nearly 75 percent of border agencies are small. One in four tribal agencies is located in a rural area, and one in three tribal agencies is located in a border area. Therefore, it is not surprising that some of our agency interviewees fell into more than one sector category.

We relied primarily on the CSLLEA to identify SRTB agencies. A small agency is one reporting fewer than 25 full-time sworn officers. A rural agency is one (excluding state police) located in a county coded between 7 and 10 on the USDA Economic Research Service Rural–Urban Continuum. These encompass urban communities of fewer than 20,000 residents that are nonadjacent to a metropolitan area and rural communities of fewer than 25,000 residents regardless of their distance from a metropolitan area. Similarly, a border agency is one located in a county within 100 miles from the U.S. border with Canada or Mexico. Finally, a tribal agency is one self-identified in the CSLLEA.

Court Sample
To identify SRTB courts, we first scanned available databases on courts to determine whether a suitable sampling frame already existed. For example, BJS periodically enumerates state and local police departments with the CSLLEA and local jails with the Census of Jail Facilities. Unfortunately, we could not identify a comparable data-collection effort for courts. For example, the Census of State Court Organization is organized by state and does not include an enumeration of courts at the local level.

To address this gap, we used an array of strategies to identify an appropriate sampling frame. For *rural* and *border* courts, we drew a random sample of rural and border counties, using the definitions outlined in Chapter One, and then identified courts located within those counties by searching the web or identifying courts in *Directory of State Court Clerks and County Courthouses: 2014 Edition* (Leadership Directories, 2013).

To identify *small* courts, we drew a random sample of offices serving jurisdictions of fewer than 100,000 in the 2007 National Census of State Court Prosecutors and then identified the

---
[3] Results for the 2014 CSLLEA had not yet been released at the time this research was conducted.

court serving the corresponding jurisdiction. Although there is incongruity between prosecutor offices and courts, we found that the courts we ultimately interviewed fit our definition of *small*.

Finally, to identify *tribal* courts, we contacted courts listed on the National Directory of Tribal Justice Systems, an online database that NAICJA hosts. Although we initially attempted to sample courts, we ultimately contacted all of them in our goal of completing eight to ten interviews with tribal court representatives.

The 35 interviews included 15 court administrators; 14 clerks; six court managers, directors, or supervisors; two judges; one attorney; one training court assistant; and two subject-matter experts. (Three courts provided more than one respondent.) One or more of four trained interviewers conducted each interview. Although we classified courts into SRTB groups, many courts qualified for more than one of these categories, requiring discretion to establish their final classification. Notably, all but two (border) courts qualified as small. Tables 2.5 and 2.6 summarize information about the court interview sample.

Corrections Sample

To identify SRTB jails, we used AJA's March 2015 jail directory (AJA, 2015) and *Census of Jail Facilities, 2006*. The AJA directory contains information on more than 3,000 jails in the United States—in most cases, a main and a secondary contact. Many jails met criteria for more than one classification as small, rural, tribal, or border; researchers randomly assigned jails meeting multiple criteria to one main category for sampling purposes.

For *rural and border* jails, we used the AJA directory's jail location information to identify those jails located in rural and border counties, as defined earlier in this report. Just over 1,000 jails were located in rural counties and approximately 275 in border counties. A *small* jail was one defined as having 50 or fewer beds; using *Census of Jail Facilities, 2006*, we identified 363 jails that fit this criterion and were not already classified as border or rural jails. We identified a tribal jail from the AJA directory as one located on tribal land or run by a tribe; about 70 jails fit this definition.

For each category of jails (SRTB), we contacted a random sample of people via email (when available) or phone to request interviews. In total, we interviewed 41 representatives from institutional corrections agencies, and each interviewee was a sheriff with oversight of the jail facility, a deputy in charge of the jail facility, or a jail administrator.

Creating a community corrections sample proved challenging because we know of no comprehensive guide or list of community corrections agencies that we could use to create a sampling frame. In addition, we had no working definition of a small community corrections agency. Instead, we searched for information on community corrections agencies in counties identified as rural or border or under tribal jurisdiction. The sampling method was biased toward states with more-complete information available online about community corrections. We contacted 32 agencies in border counties and interviewed representatives from ten, and we contacted 159 rural agencies and interviewed 26 people. We contacted 20 tribal community corrections agencies, but none agreed to participate, although we did interview one representative from a rural agency that also supervised a tribal population. We interviewed 36 community corrections representatives in total, each of whom was the chief probation or parole officer or the probation office director for that agency.

Advisory Panel

The goal of the advisory panel was to get feedback on the results of the interviews and to prioritize the needs that interviewees identified. The advisory panel brought together 34 people representing law enforcement, courts, institutional corrections, and community corrections agencies across all SRTB sectors. Five (nonpracticing) experts also joined the advisory panel meeting to provide their input. We identified panelists by selecting interviewees who had expressed interest and had subject-matter expertise in SRTB courts, law enforcement agencies, community corrections, or institutional corrections and invited representatives of the previous SRTB Regional Center's board and staff of both the Center for Rural Development and NIJ. Appendix C lists the advisory panel members.

We invited advisory board members to RAND's Arlington, Virginia, office. Participants attended sessions over a day and a half that provided background on JIC; presented the results of the interviews; and allowed for breakout sessions for law enforcement, courts, community corrections, and institutional corrections participants to review and discuss the interview findings. We asked participants to rank the identified needs. Chapter Five provides more information on the advisory panel. Table 2.7 shows the breakdown of participants.

Table 2.7
SRTB Advisory Panel Participants

Agency or Sector Type	Participants
Agency type	
Law enforcement	10
Court	11
Institutional corrections	6
Community corrections	7
Sector type (can overlap)	
Small	13
Rural	13
Tribal	8
Border	10

NOTE: Twenty-nine of the 34 expert participants were affiliated with law enforcement, court, or corrections agencies. Another five are former employees of such agencies or experts affiliated with research institutions.

CHAPTER THREE

Literature Review

In this chapter, we briefly summarize key findings from the literature on SRTB agencies. We first discuss evidence on general challenges that SRTB agencies face and then discuss literature on the needs and challenges associated with the three types of SRTB agencies of interest in this study: law enforcement, courts, and corrections. Although this report is focused specifically on the needs of SRTB agencies, we also highlight specific technologies that might be relevant to SRTB agencies and discuss their application across SRTB agencies, where known. In brief, the review highlights the limited attention that researchers paid to SRTB agencies and the need for more information on technology use in these areas.

General Challenges Facing SRTB Agencies

Criminal justice agencies in SRTB communities account for nearly 90 percent of all agencies nationwide, but relatively little research is conducted in these settings for a variety of reasons. In this section, we describe the general challenges in these underresearched and underresourced settings; however, it should be noted that there are likely substantial differences across sectors (SRTB) and justice agency type (law enforcement, courts, and corrections) that deserve careful consideration.

SRTB agencies face many public safety issues similar to those found in urban areas, as well as other issues unique to rural areas. Weisheit, Falcone, and Wells, 1994, reports that rural criminal justice agencies face crimes exported from urban areas, including gang activity and drug use, as well as crimes related to agriculture and wildlife. The Federal Bureau of Investigation's (FBI's) Uniform Crime Reporting and the National Crime Victimization Survey show distinct differences in victimization rates in urban and rural areas (Uniform Crime Reporting Program, undated; Bureau of Justice Statistics, undated). Urban areas have higher rates of violent crime and victimization: Reported crime rates are nearly four times higher in metropolitan areas than in nonmetropolitan areas, and victimization rates are approximately 55 percent higher in urban areas than in rural areas. However, sexual assaults are reported to law enforcement at higher rates in small cities and suburban areas than in urban areas, by approximately 20 percent.

Rural criminal justice agencies face unique challenges in addressing crime as well. These challenges include long travel distances for law enforcement response, community supervision appointments, and court attendance; low case volumes, which incentivize multicounty and regionalized systems to work with limited local budgets and increase efficiency; limited staff and infrastructure, which require staff to perform multiple roles; and lack of local community-based experts and resources (Vetter and Clark, 2013). Moreover, because administrative staffs and budgets in SRTB agencies are typically small, training can be cost-prohibitive because of high travel expenses or the time required away from duty.

Border agencies face their own set of challenges. Drug and human trafficking concerns are acutely problematic in these areas. Although the extent of human trafficking is unknown, researchers have found that law enforcement agencies typically have a difficult time recognizing victims of trafficking; this problem is exacerbated when agents lack specialized training (which is common in small and rural agencies). Further, the web of agencies responsible for enforcement of trafficking violations makes prosecution difficult (Owens et al., 2014). Monitoring and oversight of transient populations during adjudication and community corrections are also difficult, especially among immigrant populations. The need to collaborate with federal agencies performing border patrol activities related to drugs, smuggled goods, and immigration offenses heightens the complexity of enforcement.

There is little published literature on the tribal justice system. Perin, 2014, a review of Alaska's village public safety officers, highlights the challenges of rural and tribal policing and notes the role of technology in addressing those challenges. We do know that, despite declining crime rates nationwide, the most-recent evidence suggests that crime disproportionately affects tribal areas. Between 1992 and 2002, the rate of violent crime among Native Americans was twice the national average, while alcohol use among offenders was 50 percent higher, DUI rates were 40 percent higher, and liquor-law violations were 180 percent higher (Perry, 2004). The latter might be at least in part due to differences between U.S. and tribal criminal laws and procedures. In addition, the Centers for Disease Control and Prevention, using data from the 2010 National Intimate Partner and Sexual Violence Survey (see Black et al., 2011), reported that 45.9 percent and 45.3 percent, respectively, of Native American women and men experience physical violence over their lifetimes (Breiding et al., 2014). More-general victimization data indicate that the annual prevalence of violent victimization among Native Americans is approximately 10 percent; incidence of violent victimization is more than twice as high among Native Americans as in national data (Perry, 2004).

As a consequence of geographic spread and having relatively few employees, SRTB agencies have difficulty acquiring grant funding and training for new technologies and lack a centralized voice to affect policy and technological need assessment (Sale, 2010). A substantial literature exists on the use of technology in law enforcement, but these studies focus mainly on larger and more-urban settings (e.g., Gordon et al., 2012; Homeland Security Studies and Analysis Institute, 2012; IACP, 2005; Koper, Taylor, and Kubu, 2009; LaTourrette et al., 2003; LECTAC, 2009; NIJ, 2009; Police Executive Research Forum, 2011; Schwabe, Davis, and Jackson, 2001). Similar limitations exist within the literatures on corrections (e.g., Atherton and Russo, 2009; Jackson et al., 2015; LECTAC, 2009; LIS, 1995; NIJ, 2009; Scism, 2009; Zelenak and Goff, 2005) and the courts (e.g., Bean, 1999; Hill, 2004; Lambert, 2008; Gallas, 2001; Courts Task Force, 2001; NIJ, 2009). Existing studies have addressed a wide range of topics, including the following:

- information-collection tools and sensors, ranging from routine (e.g., closed-circuit television in prisons) to specialized (e.g., wiretapping or tracking devices)
- information-management tools, such as RMSs, used for everything from evidence to court case files
- analytic tools to leverage available data in investigation and decisionmaking

- communication tools used within facilities, with specific populations, and with the public
- manned and remotely piloted ground, air, and water vehicles
- weapon, tactical, and force innovations for intervention or site security
- training and protective technologies, from stab-resistant vests to online teaching tools.

The technologies discussed in previous research have been applied in many different ways across various types of criminal justice agencies to address a wide range of needs. In the next three sections of this chapter, we highlight what the existing literature reveals about the respective challenges that each of the three criminal justice agency types (law enforcement, courts, and corrections) faces. We also identify technologies that have been studied for application in the SRTB context.

Law Enforcement Technology Needs Among SRTB Agencies

With regard to SRTB agencies, most existing research on technology has focused on its use in law enforcement. Two national summits on small and rural law enforcement (Sale, 2009; Sale, 2010) focused on issues influencing small and rural law enforcement agencies. In particular, participants noted several common themes among their agencies, including the lack of representation in national policy and funding, difficulty in recruiting and retaining officers, inadequate staff and funds for training, and lack of resources to pursue grants and other funding.

Other research has examined more general differences between SRTB agencies and their large urban counterparts. Studies have found that officers in small agencies are less likely than officers in large agencies and metropolitan areas to specialize and more likely to deal with a broader array of issues and that officers in smaller agencies are more likely than officers in larger cities to be acquainted with the citizens they encounter and to maintain closer ties to citizens (Falcone, Wells, and Weisheit, 2002; Liederbach and Frank, 2003). Small and rural agencies might also deal with different kinds of incidents from those that large urban agencies handle. One study showed that rates of alcohol offenses, disorderly conduct, vandalism, traffic offenses, wildlife and agricultural crime, and domestic violence were inversely related to agency size (Kuhns, Maguire, and Cox, 2007).

Small and rural agencies tend to have lower expenditures per officer and resident, as well as higher clearance rates, than their urban counterparts (Falcone, Wells, and Weisheit, 2002). Police departments that serve fewer than 25,000 residents, for example, have salaries that range from $30,900 to $54,600, whereas officer salaries in departments serving more than 25,000 residents range from $45,700 to $75,100 (Reaves, 2015). Studies indicate that the work routines of police officers in small and rural agencies are often similar to those of officers in large agencies, with most officers engaged in similar tasks during their shifts, such as routine patrol, driving, report-writing, and other administrative tasks (Liederbach and Frank, 2003).

With regard to technology, research has shown that small and rural agencies are less likely than larger urban agencies to utilize—or perceive the utility of—more-sophisticated technologies, such as Global Positioning System (GPS) and car-mounted computer devices (Eastern Kentucky University, 2002; NIJ, 2004). This reluctance might be attributable to lack of resources for acquiring such technology, as well as technological infrastructure issues associated with these technologies. For example, radio communication failure is common in rural areas, particularly when officers have to travel to distant areas, so many rely on cell phones to communicate with dispatch and request backup units (Foreman, 2013).

We identified law enforcement needs specifically related to information technology (IT) through research conducted by the NLECTC Information and Geospatial Technologies Center of Excellence (Hollywood, Boon, et al., 2015). This effort identified high-priority needs for agencies of all sizes, not just SRTB agencies, but the findings might apply to the SRTB context. We identified three general keynotes as the highest-priority needs: improving knowledge of technology and practice within the law enforcement community; sharing and using law enforcement information both within and across agencies; and researching, developing, testing, and evaluating IT for systemwide improvements (Hollywood, Boon, et al., 2015).

Results from Law Enforcement Management and Administrative Statistics

BJS conducts the Law Enforcement Management and Administrative Statistics survey, last administered in the United States in 2013 (see BJS, 2015a). This periodic survey of a nationally representative sample of state and local law enforcement agencies covers such topics as personnel, budget, and fiscal issues, as well as use of technology and information systems. Results from that survey also shed light on the differences in technology use between SRTB agencies and larger or more-urban agencies. In general, the data show that non-SRTB agencies have the greatest access to or use of technology, followed, respectively, by border, small, rural, and tribal areas. Table 3.1 shows use of technology among urban and SRTB agencies.

More than half of all agencies use in-car video technologies (e.g., dashboard cameras). More than 75 percent of non-SRTB agencies use in-car video technology, while about 67 to 68 percent of small and rural agencies, 60 percent of border agencies, and 55 percent of tribal agencies use the technology. Surveillance technologies (e.g., fixed cameras) are the next most–commonly used video technology, found in more than 60 percent of non-SRTB agencies. SRTB agencies have slightly lower usage rates, with just over 50 percent of border agencies, just under half of rural and tribal agencies, and 43 percent of small departments employing surveillance technology.

Table 3.1
Percentage of State and Local Law Enforcement Agencies Using Various Technologies

Technology	Non-SRTB	Small	Rural	Border	Tribal
In-car video	77	67	68	60	55
Surveillance technology	63	43	48	53	48
Body-worn camera	25	34	37	28	59
License-plate reader	39	8	5	30	1
Computer in the field	90	65	46	82	37
Website	91	50	39	69	39
Social media	81	48	43	61	29
Report information to the public via website	74	32	23	48	25
Report information the to public via email or text	55	49	46	56	46

SOURCE: BJS, 2015a.
NOTE: BWC = body-worn camera. LPR = license-plate reader.

Across all agencies, BWCs are less common than other types of video technology. Surprisingly, non-SRTB agencies report lower usage than SRTB agencies, with only 25 percent of non-SRTB agencies using BWCs. Rural agencies have the second-highest rates of use, at 37 percent; tribes have the highest rate of use at 60 percent, although the sample is small. Rural is followed by 34 percent of small agencies and 28 percent of border agencies reporting the use of BWCs. Higher rates of adoption of BWCs in SRTB areas might stem from the higher frequency with which officers in these areas patrol alone, warranting additional surveillance and recording of incidents in the field. In addition, many of these agencies have, by definition, fewer officers to outfit with cameras, so the initial financial barriers to entry might be lower for smaller agencies.

LPRs are used in nearly 40 percent of non-SRTB agencies and nearly 30 percent of border agencies. However, they are infrequently used in small (8 percent), rural (5 percent), and tribal agencies (less than 1 percent). Especially in rural areas, this likely stems from the significant amount of geographic territory to cover with LPRs and the lack of a central location where collecting license-plate data might make significant contributions to investigative work.

Overall, officers in non-SRTB agencies have the greatest access to computerized information while in the field, followed by border, small, rural, and tribal departments. The most-frequently available information includes vehicle and driving records, followed by warrants, protection orders, prior calls for service at specific addresses, and criminal histories. Not surprisingly, data in nearly all of these categories are least accessible to tribal agencies, with these agencies only just having received access to federal criminal justice data in 2015.

More than 90 percent of all non-SRTB agencies have websites, and more than 80 percent use social media. Further, nearly three-quarters of non-SRTB law enforcement agencies report information to the public on their websites, and more than half report information to the public by another electronic method, such as email or text. Fewer border agencies report the same (about 70 and 60 percent using websites and social media, respectively). One-third to one-half of small, rural, and tribal agencies use websites and social media, but agencies in small and rural areas are more likely to report information to the public via email or text than on websites. Many agencies also use technology to gather data from the communities they serve: Nearly all agencies—whether SRTB or not—are more likely to receive information from the public via email or text than through their websites.

Tribal Law Enforcement

Tribal law enforcement agencies operate and are funded differently from their nontribal counterparts. We begin this section with a short description of key laws relevant to tribal law enforcement.

Public Law 83-280, 1953

Prior to 1953, sovereign tribal nations and the federal government had concurrent jurisdiction over criminal and civil matters in tribal lands, similar to the jurisdictional arrangements that states have with the federal government. With the enactment of Public Law 83-280 in 1953, Congress shifted federal jurisdiction to state governments in six states: California, Minnesota, Nebraska, Oregon, Wisconsin, and Alaska (when it became a state). Since then, other states have assumed full or partial concurrent jurisdiction in tribal lands under Pub. L. 83-280.

In states that assumed concurrent jurisdiction, coordination between state and tribal authorities has frequently been difficult or unsatisfactory, marked by jurisdictional conflicts, poor government-to-government communication or policy coordination, and confusion about mutual responsibilities and authorities (Champagne and Goldberg, 2013).

In one of the few empirical studies of tribal criminal justice in Pub. L. 83-280 states and non–Pub. L. 83-280 states, Goldberg and Champagne, 2007, the authors found that tribal agencies in Pub. L. 83-280 states receive much less federal funding, particularly BIA funding, to support tribal justice systems and therefore must rely more heavily on state criminal justice services. Indeed, many tribes in Pub. L. 83-280 states do not receive sufficient support to maintain their own justice systems, making them entirely dependent on state services.

Typically, however, residents on reservations in Pub. L. 83-280 states rate the services they receive from state law enforcement more poorly than the federal and tribal agency services in non–Pub. L. 83-280 states. According to Goldberg and Champagne, 2007, pp. vi–vii,

> state or county police serving Public Law 280 reservations are rated by Public Law 280 reservation residents as less available, slower in response time, less prone to equally attend to minor or serious calls, provide less beneficial patrolling services, less willing to act without authority, more frequently decline services owing to remoteness, and are located farther away than federal-BIA and tribal police on non–Public Law 280 reservations.

Public Law 93-638, 1975

The Indian Self-Determination and Education Assistance Act of 1975 (Pub. L. 93-638) provided a mechanism whereby tribes could take over the administration of services that, until then, the federal government had provided. Tribes with Pub. L. 93-638 agencies administer their own law enforcement, courts, and corrections systems under so-called self-determination contracts with the BIA. A second funding mechanism called a BIA compact was created that provided some tribes with block-grant funding that could be allocated across a range of approved activities, including support for justice system agencies.

For tribes in non–Pub. L. 83-280 states, BIA contracts might be available to provide basic funding for law enforcement, correction or detention, and courts. The BIA currently supports 190 law enforcement offices; 33 are administered by the BIA; and 157 are administered by tribes through self-determination contracts or compacts. The BIA provides some funding for 115 of the 300 courts that tribes administer (BIA, 2013). The BIA supports 96 detention programs, of which 25 it operates and 71 are contracted to tribes.

The BIA has not, however, contracted with tribes to administer their own law enforcement services in the Pub. L. 83-280 states, where that law shifted jurisdiction for criminal and civil matters from the federal government to the state. This creates significant differences in the administration of justice system services for tribes in and out of Pub. L. 83-280 states.

In Pub. L. 83-280 states, tribes might be unable to afford their own law enforcement services, or they might depend heavily on grant funding from DOJ or other sources to continue to operate. The effect of these arrangements is that only 21 percent of tribes in mandatory Pub. L. 83-280 states have tribal law enforcement agencies, compared with 74 percent of tribes in non–Pub. L. 83-280 states and those that have only partially implemented Pub. L. 83-280 authorities (Goldberg and Champagne, 2007).

Some tribes in non–Pub. L. 83-280 states have not contracted with the BIA to administer their justice agencies. These tribes receive law enforcement or other criminal justice services that the BIA administers.

Tribes without BIA funding for justice system services—or those who allocate their block grants to other activities—can, in some cases, fund these services with gaming revenues, taxes, or DOJ grants, although the short-term funding that grants provide makes reliance on them for core operations difficult or impractical.

There is little research on the operational needs and challenges of tribal law enforcement agencies. The research reported in Wakeling et al., 2001, employed site visits, literature reviews, and surveys of tribal policing agencies to assess the challenges that tribal law enforcement agencies face. They note that, although workloads of tribal agencies are high (and increasing in many areas), staffing, facilities, and equipment are often inadequate to meet departmental and community demands. Many agencies could not provide full coverage of police services around the clock and across the wide geographic area of the reservations. In addition, lack of documented policies and procedures was common. Some challenges might be attributable to lack of resources, with tribal agencies reporting between 55 percent and 80 percent of the funding of non–Native American agencies.

Wells and Falcone, 2008, provides one of the few quantitative examinations of tribal policing on Native American reservations. That study used data from the 2000 CSLLEA and *Census of Tribal Justice Agencies in Indian Country, 2002*. The authors noted that more than half of the 314 tribes responding to the Census of Tribal Justice Agencies in Indian Country reported having law enforcement agencies employing sworn tribal personnel with general arrest powers. The authors found large differences between tribal agencies according to what entity had primary jurisdiction for criminal matters in tribal lands, as codified in Pub. L. 83-280. For example, tribal agencies in Pub. L. 83-280 states were less likely than optional and non–Pub. L. 83-280 states to have police agencies; furthermore, when the Pub. L. 83-280 states did have tribal police agencies, those agencies were less likely to be funded through BIA contracts. As a consequence, tribal agencies in these states might experience more challenges and reduced capacity for exercising tribal authority. The authors also found that the majority of tribal police departments were established recently (during the 1980s and 1990s), and this was particularly true of those in Pub. L. 83-280 states. Most tribal agencies were small, employing fewer than ten sworn officers, and located in nonmetropolitan areas serving fewer than 20,000 residents. Although less involved in criminal investigations than nontribal agencies, their main police functions were similar to those of agencies of comparable size. However, they tended to be more involved in providing supplemental policing services (e.g., court security, jail operations, serving civil process) that reflect the unique contexts in which they operate.

Border Law Enforcement
There is a larger amount of literature on policing the U.S. borders with Mexico and Canada; however, the bulk of research in this area has focused on federal enforcement rather than local police. Some issues, such as immigration, particularly of undocumented immigrants, as well as human trafficking, might be more salient for—but are by no means restricted to—local law enforcement agencies in border areas (IACP, 2007). Increased federal law enforcement activity in border areas might affect local criminal justice agencies. For example, increases in drug seizures at the border by U.S. Customs and Border Protection agents might affect local law enforcement, courts, and corrections agencies (NLECTC, 2011). Thus, it is critical to identify the needs and challenges of local law enforcement agencies located in these areas, which are not comprehensively documented in the extant literature.

Court Technology Needs Among SRTB Agencies

Many of the themes common to SRTB law enforcement agencies also apply to SRTB courts. For example, such factors as greater geographic distance, sparse population, lack of resources (including staff) and specializations, professional isolation, and close interpersonal relationships between justice personnel and citizens all present challenges to courts, just as they do to law enforcement agencies. Challenges that are more unique to courts include the lack of programs and services to address mental health, substance use, indigent defense, and self-represented litigants, as well as the increasing need to address language barriers throughout court proceedings (Nugent-Borakove, Mahoney, and Whitcomb, 2011).

Rural out-migration, particularly of young people, might pose a particular problem for rural courts. In these communities, court budgets, staff, and caseloads are declining while the working-age population grows. In addition, consolidation of adjacent courts to ensure staff coverage often creates additional challenges because rural courthouses are geographically isolated from one another. To address some of these challenges, it might be necessary for rural courts to undergo reorganization, a process in which technology might play an important role. A recent report by the National Center for State Courts (NCSC) on the challenges confronting rural Minnesota courts includes a discussion of technology initiatives to address these challenges (NCSC, 2010). These include a statewide management system, real-time updating of court proceedings, and a centralized payment system to process citations and other transactions between the courts and citizens, all of which are expected to reduce staff time and expenses in rural courts. These challenges echo those identified as high-priority law enforcement information-technology needs (Hollywood, Boon, et al., 2015).

Another major challenge that SRTB courts face is ensuring equal access to justice. Recruitment and retention of attorneys who provide indigent defense and other legal services in rural areas is a considerable challenge. For example, in rural Arkansas, there is less than one (0.72) attorney per 1,000 residents, compared with 4.11 per 1,000 across the nation (Pruitt et al., 2015). This is one area in which technology has been used to address the challenge. Recently, the American Bar Association (ABA) provided a grant to the Wyoming Access to Justice Commission to place remote-access sites—from which pro bono attorneys can meet remotely with clients via Skype—in five underserved rural communities (Wyoming Center for Legal Aid, 2013). In addition, the ABA has provided a guide to delivering legal services in rural areas (ABA, 2003) and specifically recommended use of technology to overcome barriers in providing legal aid to rural residents. Recommended projects include development of Internet clearinghouses with legal information for attorneys, advocates, and clients, as well as videoconferencing to reach rural clients.

Although there is less information on SRTB court technologies, a growing number of efforts address the challenges that rural courts face. Recent efforts to analyze and solve problems in rural courts have been undertaken by the Nevada Commission on Rural Courts, the Justice Management Institute, and the National Association for Court Management (Mahoney, Carlson, and Baehler, 2006). In 2008, BJA also funded the Rural Court Information Network initiative (Justice Management Institute, undated) to provide a series of seminars in order to identify areas of need and technological and other innovative strategies to address these concerns. Court leaders (judges and administrators) from eight states led the seminars.

NCSC also identified areas in which technology has been applied in rural courts. Such technology includes videoconferencing to address the distance that defendants might have to travel to court; web-based learning opportunities to train court personnel; and the use of online systems to address citations, tickets, and fines instead of traveling to a court office (see Nugent-Borakove, Mahoney, and Whitcomb, 2011). NCSC also maintains both "Rural Courts Resource Guide" and "Technology in the Courts Resource Guide" (NCSC, undated [a]; NCSC, undated [b]); however, neither of these resources addresses the specific technology needs of SRTB courts.

Although each of these initiatives is using innovative technology to solve unique needs in rural areas, none has been evaluated, so their effectiveness has yet to be established.

Tribal courts might face challenges similar to those that other small and rural courts face, as well as additional challenges, such as jurisdictional confusion, lack of communication, and misunderstanding between state and federal authorities (Arnold, Reckless, and Wolf, 2011; Arnold, Reckess, and Wolf, 2012). Several strategies have been utilized to improve communication, including tribal and state court forums; joint-jurisdiction courts; and written cooperative agreements between tribal, state, and federal governments. Although less information is available on technologies in tribal courts, some resources focus on tribal issues, including the NCSC's "Tribal Courts: Resource Guide" (NCSC, undated [c]), the tribal justice section of the Center for Court Innovation's website (Center for Court Innovation, undated), the Tribal Court Clearinghouse of the Tribal Law and Policy Institute (Gardner, undated), the National Indian Law Library (Native American Rights Fund, undated), and NAICJA (NAICJA, undated). Several of these organizations have ongoing projects in tribal courts, some of which are related to technology (NCSC, 2009), but the majority focus on implementing culturally competent tribal practices, such as traditional dispute resolution (Metoui, 2007), wellness (tribal drug) courts (Tribal Law and Policy Institute, 2003), and tribal court-appointed special advocate programs (Frey, 2002).

Like with law enforcement, there is not a clear body of literature on the challenges and needs that courts located in areas near the border face or on technological solutions to address these needs. Such issues as language barriers might be salient to border courts, although these issues are by no means restricted to border areas. And, like on law enforcement in border areas, more research is needed on courts located in these jurisdictions, and evaluations are needed to understand the effectiveness of technologies.

Corrections Technology Needs Among SRTB Agencies

Little research has focused specifically on corrections technology for SRTB agencies, whether in prisons or jails or for community corrections contexts (Ruddell and Mays, 2006). Much of the research on rural corrections agencies has been concerned with decisions on prison location and the use of prisons as a source of economic development in rural areas (King, Mauer, and Huling, 2003). Most of the research related to technology concerns large or urban corrections agencies. However, lessons from these studies can help SRTB agencies make decisions on the adoption of new technology.

Current research at RAND, conducted under NLECTC, aimed to identify the highest-priority technology needs for criminal justice agencies (Jackson et al., 2015). The RAND study identified 19 areas, or needs, as the highest priorities for *community* corrections agencies, all of which fell into one of two categories: (1) information and communications or doctrine, tactics, and management or (2) behavioral knowledge development and training. Echoing the needs of law enforcement agencies and courts, many of the information and communications needs focused on improving "fragmented" systems of information-sharing among siloed agencies and improved data extraction and usability by a range of users. In the doctrine category, the authors identified improved training and training materials for handling offenders with mental health problems and appropriately sanctioning probation and parole violations (Jackson et al., 2015).

The authors identified 29 high-priority needs for *institutional* corrections agencies, which fell under the categories of (1) facility operations and population services; (2) information and communications; and (3) doctrine, tactics, management, and behavioral knowledge development and training. Issues with detecting and interdicting contraband from a range of sources and methods dominated the needs identified, as did managing inmate access to and use of technology, including Internet and communication technologies. Also noted as challenging were the collection, monitoring, review, and storage of the enormous amounts of data produced through video, phone recording, and other technology systems used in the facilities (Jackson et al., 2015).

Although the needs identified for both community and institutional corrections apply to the SRTB context, and significant research has been conducted on some of the technologies that could be used to address these needs, scant work has considered the implications of these needs or the technologies that might address the needs specifically among SRTB agencies. Applegate and Sitren, 2008, a comparison of rural and urban jails, provides some insight into the needs of rural jails and areas that might be addressed through technology. They found that, although staff-to-inmate ratios were lower in rural jails, rural jails were less likely to test inmates for various medical conditions, including HIV and acquired immunodeficiency syndrome (AIDS), and that rural jails were less equipped than jails serving urban areas to provide services, such as medical services, employment training, or counseling. The authors noted, however, that rural and urban jails face many of the same challenges, suggesting that similar technologies might be applied in both contexts to address shared issues.

Kellar, 2001, a review of challenges that jails in Texas face, emphasizes the need to take jail size into account in any analysis of jail operations. That work identified a variety of issues relevant to small jails that were similar to those that prior work noted. The challenges identified in Texas included the high costs of providing medical services to inmates; the need to manage offenders with mental health issues; and the difficulty of providing other types of services, including substance-use treatment, education, and employment training. Kellar, 2001, also identifies the three most-pressing needs of jail administrators: medical service provisions, budget restraints, and employee training. More than a decade later, in Jackson et al., 2015, the authors' identification of high-priority needs echoed these and suggests that significant efforts to address those needs have not been successful.

Conclusion

The current review has highlighted the dearth of information on the needs of SRTB agencies and the technologies that might be applied to address these challenges in the SRTB context. Researchers have paid limited attention both to identifying and assessing needs and to identifying technology to assist SRTB agencies. We also find that there is even less information available measuring the *effects* of these technologies across the three criminal justice agency types, including the value that technology might provide through increased effectiveness, the more-efficient use of resources, and other outcomes, such as officer and employee safety.[1] And among all agencies—large ones included—technology has not always been used in the most effective or efficient manner to improve policing practice (Koper, Lum, and Willis, 2014).

To prioritize the acquisition of technologies for use in SRTB criminal justice agencies, it is critical to understand the ways in which technologies are used, the outputs and outcomes of their use, and the ways in which they might be used to address unmet needs. It is also essential to understand how technologies might be matched with the unique needs of the SRTB justice systems, which might, in some cases, be very different from those in larger or more-urban environments.

Selecting clear and defensible technology priorities has been a challenge in the field of organizational technology planning, arising from the uncertainty associated with predicting technology developments and the ways in which the adoption of technologies affects organizations and societies, as well as the difficulty in establishing a common basis for comparing technology's often vastly different effects on various users and sectors. Such a comparison has been made for criminal justice and broader emergency responder communities (e.g., Garwin, Pollard, and Tuohy, 2004; Gordon et al., 2012; IACP, 2005; Koper, Taylor, and Kubu, 2009; LaTourrette et al., 2003; LECTAC, 2009; Courts Task Force, 2001; NIJ, 2009; Royal, Donahue, and Kirby, 2008) but has never focused specifically on SRTB settings. The current review suggests that more attention needs to be paid to the needs of SRTB agencies, the use of technology to meet those needs, and the effects of technologies when applied in an SRTB context.

---

[1] For further discussion of this point, see Chapter One.

CHAPTER FOUR

Agency Interviews

Like we have demonstrated in the prior chapters, despite the fact that the majority (90 percent) of criminal justice agencies in the United States fit at least one of our SRTB criteria, research into the needs of these agencies and effective solutions to meet those needs is sorely lacking (NIJ, 2004). To address this gap in understanding, the JIC team conducted interviews with SRTB practitioners in all three agency types. Across these agency types, our interviews included representatives from 50 border agencies (12 law enforcement agencies, eight courts, and 30 corrections agencies) and 27 interviews with representatives from tribal agencies (nine law enforcement agencies, nine courts, eight institutional corrections agencies, and one community corrections agency). Note that some agencies are considered both border and tribal.

Across 25 interviews with tribal agencies, as well as interviews with agencies bordering tribal nations and the BIA, we identified a range of operational needs that might be common among tribal justice agencies. That said, our small sample of tribal agencies was also quite diverse. A significant source of variation results from federal laws covering governance of tribal lands that apply differently by state and tribe, as discussed in Chapter Three.

There are two categories of corrections facilities on tribal lands. The first group includes facilities that the BIA wholly funds, staffs, and operates. The other group, commonly referred to as 638 facilities (after Pub. L. 93-638, 1975, which set up the type of division of labor managing them), receives base funding from the BIA, while the local tribal government prepares the budget and manages operations. We interviewed seven administrators or wardens, as well as a BIA district-level official. Most interviews we conducted were with people from 638 facilities, which operate with a higher degree of autonomy than bureau facilities.

Chapter Two provides more information about the interview samples.

Many common themes emerged from these interviews, including challenges and barriers related to geography, crime, funding, personnel, infrastructure, IT (both in capacity and implementation), data-sharing and communication, and legal and policy issues. Within each category, we highlight findings relevant to each type of agency, including those that apply to border and tribal agencies.

We organize the rest of this chapter by theme. We identified themes using open-ended questions, and interviewees raised issues that they felt were most important. Because we did not prompt respondents about specific themes, some respondents do not discuss particular themes. However, the lack of discussion of a theme in relation to a particular agency type does not mean that the issue is not relevant to that agency or agency type.

Geographic Challenges

Many respondents across agencies mentioned challenges associated with geographic isolation and challenging topography. Specific challenges for courts and corrections facilities included difficulty in getting jurors and defendants to courthouses on trial dates because of long travel times and a lack of transportation resources in the community. Geography also increased difficulty in combating drug cartels with superior technological capabilities. Geographic isolation creates staffing problems because of transportation times for police officers responding to calls, community corrections and court officers conducting community supervision, and jail staff transporting inmates for external health services and court appearances.

Law Enforcement

Multiple police administrators and managers from SRTB agencies reported significant operational challenges associated with geographic isolation. Several agencies indicated that their location isolated them from important infrastructures and services that they could not provide on their own (including juvenile detention for one facility and jail for another). For agencies lacking key services, it could take several hours to transport arrestees to their destinations. For most SRTB agencies, such arrangements place tremendous strain on the department because their staffing and resources are already limited. As one officer stated, "You have to drive the person three hours there and three hours back. That cuts into your staffing, so coverage is an issue."

Border agencies often highlighted the need to transport arrestees long distances. One border agency has been using video for initial court appearances and attorney/client meetings, which has worked well to save money and time, but the district court does not allow video for sentencing or trial sessions.

Geography and topography can also affect law enforcement agencies' ability to implement technology. One tribal police chief explained that his department invested in gunshot detection technology—devices that can detect the location of gunfire—but deployment in his community was rendered difficult because the community lacked tall buildings on which to place the sensors for accurate detection. The same agency also had difficulty with its crime analysis software because the jurisdiction used three different address schemes, which made locating 911 callers and mapping crimes difficult. Each tribal resident was supposed to provide a physical address for 911 calls, but many of them did not know that: "You could talk to people in the community, and they don't know what their address is." Another police chief, leading an agency based in the South, mentioned that any technology would need to withstand heat in excess of 100º F all year long. One police chief explained that bad road conditions and the remote nature of his area could lead to difficulty in getting external help when needed. In sum, the majority of border and rural interviewees expressed at least some concerns about lengthy response times when their officers need backup.

Some agencies also indicated that, because of their expansive geographic jurisdiction, they did not receive radio and cell phone coverage in certain areas when out in the field. As one officer stated, "Our radios cover 70 percent to 75 percent of the area we are responsible for . . . causing huge gaps for not just officer but public safety."

Courts

Almost all courts, especially rural courts, emphasized the impact of geographic isolation. Respondents noted that lengthy travel times limit timely access to services, such as IT, training, client treatment, and reentry services. In some cases, judges, court reporters, lawyers, or jurors might be regularly required to travel hundreds or even thousands of miles to meet their appointments. More than one court representative pointed out that there is no public transportation to offset these challenges in that county, which disproportionately affects the elderly, those with license revocations (e.g., DUI probationers), and children who are involved in court proceedings. The ABA's report (ABA, 2003) on legal services in rural areas identified these issues more than a decade ago; the same issues still present challenges to residents in rural areas, according to the JIC interviewees.

With the exception of some larger jurisdictions (e.g., San Diego, El Paso), most counties along the border are rural areas with populations that might be spread out across a large geographic area, with some legal residents even residing across the border. This can pose additional challenges for various parties (e.g., jurors, plaintiffs, defendants) who are required to attend court. For example, one court administrator noted,

> Jurors are 80 or 90 miles from [the city], so may have to travel 200 miles in [the] same county. On the border, it's the same way. Both counties are very rural counties, but neither has public transit, so we run into that problem of getting jurors, defendants, various parties to courthouses. Also, because they are on the on border, if you cross the bridge at [the city], there is a fee to cross the bridge at the border crossing.

Tribal court respondents also noted geographic challenges. A couple of tribal courts highlighted the difficulty of conducting community supervision over long distances, resulting in offenders being either unsupervised or separated from their families in order to receive supervision. One respondent noted that, if the probation department knows that an offender's home community is "100 miles from the closest agent, [the offender] may get denied [probation]. It can have an effect on [the offender's] ability to go back to [the] community and be with [a] positive support system." The long distances also delay the implementation of operational changes to outlying satellite offices. Respondents suggested the use of technology to overcome barriers to serving rural clients, including Internet clearinghouses and videoconferencing. For example, in Wyoming, the ABA is funding the implementation of Skype to support meeting with and serving rural clients remotely over video (Wyoming Center for Legal Aid, 2013).

Geography can also affect an agency's access to interpreter services. Some courts, particularly those near the Mexican border, expressed concern about the lack of adequate interpreter and translation services for the increasing number of non–English-speaking clients. This makes it difficult for such clients to get adequate access to justice. As one respondent noted, "That's something we have to address even more than is currently being done because, otherwise, people with limited English proficiency will have less access to courts than they're entitled to." Respondents noted that technology might be used to provide interpreter services in remote areas. Nugent-Borakove, Mahoney, and Whitcomb, 2011, identifies language differences and local service provision as barriers to justice for court systems, especially those in rural areas.

Institutional Corrections

The remote locations of many SRTB facilities also present challenges for many jail operations. Like with other types of agencies, geographic isolation can limit access to potential staff, services, community interaction, and visitation. Many smaller remote facilities also lack opportunities to take advantage of economies of scale, which might be necessary to acquire technology solutions requiring significant up-front investment or that are affordable only if purchased with volume discounts. One respondent suggested a potential solution through pooled contracts:

> We recently switched our canteen vendor. We discovered [that it was helpful] for inmates' families to be able to provide them canteen services, . . . but, in another county smaller than us, the vendor won't even look at them because the revenue isn't high enough. . . . If we could do things more jointly, [both counties could benefit].

To address this issue, some agencies reported, they work with neighboring jurisdictions to create buying collaboratives or to make cooperative purchases, allowing participating agencies to take advantage of economies of scale. Interviewees made the point that increasing the purchasing power of small and rural agencies could help reduce their costs and put more-expensive technologies in their reach.

Community Corrections

Community corrections agencies noted that geography affects staff's ability to conduct home or work visits. In agencies that required home visits—especially among officers with higher-risk caseloads—getting to clients and getting clients to services posed significant challenges. This was especially so in states where community corrections are handled at the state level and supervision districts contain multiple counties. One agency struggling with this issue noted that a GPS device that optimizes trips between multiple locations is very helpful and allows officers to visit more offenders in one day than when personnel planned the trips manually. Respondents also noted a lack of public transportation and pointed out the high costs associated with using transportation other than a personal vehicle to attend sessions or meetings in remote agency offices.

Some agencies use video links with courts and jails to conduct pretrial arraignments and presentencing investigations (PSIs) to minimize time spent on transporting offenders over long distances. Additionally, one agency providing video treatment options had to get judicial buy-in for that element of supervision. Another agency reported trying to set up a video link with the jail to make PSIs easier but being stymied by firewalls in the jail system and finally abandoning the project.

Funding Challenges

Interview respondents universally mentioned funding shortfalls and that they affect many of the other challenges that multiple sectors and agency types face. Small budgets could lead to recruiting, retention, and training challenges. Like with geographic isolation, resource constraints also affect agencies' ability to do their jobs efficiently and effectively.

Law Enforcement

A recurring theme among SRTB law enforcement agencies revolves around insufficient funding. Many respondents noted that their budgets have been decreasing over time, and, because of financial hardship, several police departments reported that they could not hire enough officers to provide the community with essential services. As one officer stated,

> We are going backwards. We are a state that [has some of the highest rates in] the nation in domestic violence, assault, sexual assault, and death by accident. And our budget is being cut barebones. We cannot meet the demands for services.

For border agencies, interviewees frequently mentioned challenges stemming from limited budgets as a priority issue. Two police chiefs from border areas highlighted the importance of the U.S. Department of Homeland Security's (DHS's) Operation Stonegarden grants for the financing of border law enforcement agencies and the difficulties border agencies might face in the event that these grants are not available. One interviewee also lamented recent substantial cuts to available DHS funding, which, in his perspective, have negatively affected border law enforcement agencies:

> In 2011, DHS defunded half of the border areas—originally, DHS funding was $2.6 billion per year; within two years, it was down to $900 million. The problem with that is that, since 9/11, we began building capabilities, across the nation, but specifically at the border, [we built] things like preventive radiological nuclear detection, rapid-response teams, reverse 911 systems, emergency mass communication capabilities. As those projects were defunded, we lost the ability to maintain and sustain the capabilities that were developed since 9/11.

One chief from a border area noted that his agency, unlike its larger counterparts, could not receive bulk discounts when purchasing equipment, further adding strain on limited finances. Another chief from a border area mentioned that recent limits that the attorney general placed on his program hamper his ability to use asset-forfeiture funds to help finance his agency: "Money laundering and drug smuggling . . . [are] another challenge. We used to be able to confiscate this money. However, the attorney general has abandoned the federal asset-forfeiture program, which means we cannot process and obtain this money."

The vast majority of tribal law enforcement interviewees mentioned financial constraints, which invariably translated into staffing, technological, or operational challenges for their agencies. In five instances, interviewees noted that the limited amount of available funding poses an issue, while two police chiefs whose agencies relied exclusively on grants as a source of funding expressed concerns about long-term sustainability and ability to plan their operations. For instance, one interviewee specifically highlighted the cost associated with acquiring and maintaining compatibility and interoperability of systems as an important factor.

One tribal police chief explained that, because his state does not receive BIA funding because of Pub. L. 83-280, he needs to rely on grant funding. If the agency were to run out of federal funding, it would have to rely on the tribe for funding. Wakeling et al., 2001, a review of tribal law enforcement agencies, also identifies extreme resource limitations, reporting that tribal agencies have to provide law enforcement services with less than half of the funding of nontribal agencies. The funding shortage prevents some agencies from maintaining their existing technology and from acquiring new technology, especially given the high cost of much technology.

To address the lack of funding, many SRTB law enforcement agencies stated that they relied heavily on grants, even though several of the officers expressed skepticism about smaller agencies' ability to beat out larger agencies for grant funding. As one officer stated,

> One of the biggest hurdles is funding to be able to obtain and maintain technology. I'm guessing, out of the 50,000 cameras that the president has said to issue out, 0 percent will come to any agencies that are 50 sworn or less. They will all go to larger agencies.

Courts

SRTB courts also reported challenges associated with budget constraints and instability. Some respondents pointed to broad socioeconomic causes for funding deficits, such as the national recession or unreliable state tax revenues. As one respondent explained,

> Courts haven't recovered from the recession. We're resource-starved, don't have a very supportive administration. We've lost tens of millions of dollars. It's one thing to have adequate technology support; sometimes you need enough boots on the ground to make that transition. Our workforce is 20 percent smaller than it was eight years ago. We're doing it on the cheap; we rob from one part of the organization to staff another part of the organization, [and have] backlogs that affect public access and service.

Other resource constraints were common as well. Many courts have experienced increased caseloads, which some respondents believe to be due to changes in arrest priorities and to an increase in self-represented litigants. Small courts also lack the resources to invest in sophisticated technologies, particularly ones that become cost-effective only with a large number of licenses.

Respondents explained that smaller courts are often limited to a single judge, or even worse, if no single judge is available, multiple judges from different courts might have to preside over a single case, reducing case continuity. A respondent stated,

> If you have a person in that part-time position and they only do a probate case once every three years, it is difficult for them to get accustomed to all the different things, because they are expected to run the gamut of the court system. In rural areas, you can't specialize like in urban courts.

Tribal courts also described funding problems. One clerk mentioned that she has seen budget cuts of 10 to 15 percent each year for the past three years. Several tribal court clerks noted that they depend on external grants for training, equipment, and even personnel. A clerk explained that, if they "don't get funding from an outside agency, that position disappears." Another tribal court clerk noted that she spends most of her time writing grant proposals, taking her away from other urgent court needs.

Institutional Corrections

Despite the varied composition of the sample, institutional corrections respondents consistently mentioned issues related to budgeting and staffing. Funding often lags behind the need, leading to low wages and slow adoption of new technology and, in turn, difficulties attracting and retaining qualified staff. Smaller agencies also reported a disadvantage in applying for state and federal grants because they lack the time and resources to devote to applications.

Concerning budget, respondents from tribal agencies generally felt that the total levels of funding are insufficient. As mentioned earlier, 638 facilities receive base funding from the BIA. They also receive a portion of their overall budgets from their governing tribal bodies. One tribe in particular informed us that the portion of funding that its facility receives from the tribe was being cut. And, although the number of inmates housed in the facility was not changing, the facility would have to continue operations at the same level with reduced funding.

Respondents from institutional corrections emphasized the lack of resources to address inmates' medical and mental health problems. Substance use, chemical dependency, and mental health are common issues that corrections agencies face, but service provision often requires specialized staffing, equipment, and funding to support them. These findings provide more evidence to Applegate and Sitren, 2008, that rural jails were less equipped to test and treat inmates for various medical conditions, including HIV and acquired immunodeficiency syndrome (AIDS).

Community Corrections

Community corrections agencies reported similar issues with funding. Not surprisingly, nearly all interviewees reported financial constraints that kept them from implementing all of the technology upgrades or desired improvements. Agencies reported getting assistance from a wide variety of funding sources, including the state, the locality, and grants—and often a combination of all three.

Limited resources at the state or local agency level lead many agencies to pass costs on to offenders—through supervision fees or charges for certain types of supervision. Although offenders on supervision tend to be worse off than nonoffenders in rural and urban areas alike, and although supervision fees are also used in urban areas, their impact on offenders can be more acute in rural areas. For example, as described earlier, gaining access to affordable transportation is more difficult in rural areas, where services might require long daily trips, which can be cost-prohibitive to offenders without their own cars or driver's licenses. Alternatives to in-person services, such as online classes, often require up-front payment on a credit card, which excludes many offenders.

Social Service Provision Challenges

Providing needed social services to the populations they serve is a major challenge for all communities of practice. Respondents from courts, institutional corrections, and community corrections noted a range of technology-related challenges that affect service provision, especially related to mental health and substance-use treatment, and the issues appeared to affect people who might come into contact with all elements of the justice system.

Courts

Several court respondents noted that unfavorable economies of scale disadvantage SRTB courts. Specifically, small courts cannot benefit from the specialization of judges and courts (e.g., problem-solving courts) because larger, well-resourced establishments are typically necessary to justify the investment for such specializations. As one respondent explained,

> It's extremely difficult for a very small court in a rural county without a lot of resources to implement the best practices and conform to the required program fidelity to be effective. Drug courts, veterans courts, mental health courts. . . . The treatment resources aren't available, the judges, because they have to cover all kinds of cases, and they don't have high case frequency. All these things mitigate [sic] against expertise, best practices, and economies of scale that make it doable.

The lack of specialty courts can take away opportunities for people to avoid detention and to receive needed treatment.

Institutional Corrections

The current methods of treatment in jails as institutional corrections interviewees described them range from having social workers or nurses on staff, having part-time nurse coverage, or outsourcing to nearby external facilities. The latter requires transportation and coordination of inmate needs with staff availability. It also limits corrections facilities' ability to respond to inmates in crisis:

> There is a [mental health] bed shortage. . . . In the end, everybody seems to end up in the county jail, which is unfortunate. It costs $1,100 per day in a mental health facility and $55 per day in a county jail. So it's not hard to see why they get pushed to county jail. They place them in a holding cell with a camera in it.

Kellar, 2001, a review of Texas jail operations, reports the same challenges, with mental health treatment needs high and jails' ability to provide such services low. More than a decade after Kellar's study, administrators in small, rural, and border jails were reporting similar, if not worsened, conditions.

For example, tight budgets and remote locations limit the border corrections population's access to mental health, medical, and substance-use services. Several agencies cited a lack of local capacity for substance and chemical dependence programs, as well as mental health programs. Many border agencies reported a lack of social services in their communities, and the cost of adopting telemedicine and remote health provision technology is a barrier. One jail administrator in a border region summarized the problem as follows:

> We don't have vendors come in and help us provide services with people we work with for mental health, employment, transportation, chemical dependency—all these things—to get people back on their feet in prosocial fashion. We have people driving 50, sometimes 60 miles to get services that they need, like mental health [treatment].

Many institutional corrections agencies reported interest in telemedicine and remote treatment services but noted large start-up costs, financial and otherwise, that can hinder implementation. Many facilities lack the physical space and funding to purchase the equipment required to run that type of program. Facilities also have to locate a doctor or medical system—in some cases, the local hospital system—that is set up to conduct telemedicine. Coordinating with providers and insurance companies can be a huge barrier to implementing telemedicine: "The biggest challenge is working out the details. The technology is there. There are companies that do this, but not [many do it] for corrections agencies."

Respondents also noted that some technologies cannot be implemented easily because they require on-site personnel for implementation. For example, some telemedicine services—particularly those *not* focused on mental health—require a trained nurse to be with the inmate while the video session is happening in order to conduct any tests or simple medical procedures as the doctor directs. Thus, to use the technology, a facility needs to coordinate telemedicine schedules with an on-site nurse. For this reason, some facilities have chosen to conduct only mental health services remotely over video. However, regardless of these challenges, those that have implemented telemedicine services report being pleased with the benefits, particularly the opportunity to have a mental health professional on call to work with inmates in crisis, instead of having to wait until business hours for assistance.

One respondent pointed to a potentially more systemic problem in that, even when mental health issues are addressed in their facility, the successful routine that was established inside ends when an inmate is released: "The problem with mental health issues and inmates is they will come in to the facility, get back on their meds, get stabilized, and be released because the offenses are small." After release, if no aftercare or reentry support services are available, the inmate starts the cycle over. This corresponds to other agencies reporting a lack of broader social services:

> We have 24 hours of contracted mental health services [in the facility per week.] Ultimately, it fails because, although we do the prep work here, once they leave the jail, we cannot require [former participants] to take [their] medication or make appointments.

Although such problems are not unique to SRTB agencies, the resulting challenges might be greater because of scarce or dispersed resources available to provide care.

Community Corrections
Community corrections personnel reported difficulties in getting their clients into services, whether because of the high cost of the program that the probationer has to pay, the lack of programs available in rural areas, or the lack of openings in existing programs. Indeed, one interviewee described issues with getting clients into treatment as "multifaceted": "[We] are doing a great job of identifying people at arrest who have significant issues, but one of the biggest challenges is having easy access to necessary services."

Other agencies reported an inconsistent flow of clients with specific treatment demands, making it hard to lobby for more service provision in their areas: "We have ten clients with the same needs one week, and a couple of months later, we'll have no clients with those needs." A community corrections interviewee working in a border region noted, "We have seasonal employees moving in and out, and trying to provide services to them is a challenge."

Many interviewees recognized that technology can offer solutions to these problems. One agency identified online programs that probationers could use to fulfill treatment requirements, while several other agencies conducted group therapy sessions and client meetings using remote video links between agency offices. For example, some probation officers are trained to run group therapy sessions so the agency can deliver services itself when they are not available elsewhere. Combined with remote video linkage, this approach can provide services to more remote clients. Some interviewees also reported that specialized service provision and treatment, such as anger-management treatment or shoplifting programs, are difficult to find and difficult to place clients into. Online or remote options could offer more access to these less common treatment types as well. Although these options might not address more-serious issues, such as drug or alcohol treatment, they might free up probation officers to focus on clients with higher need levels.

Personnel Challenges

Budget challenges and geographic isolation can contribute to problems with staff recruitment, retention, and training. Constrained budgets leave little money available for hiring or training, and salaries at SRTB agencies are often not competitive with those of larger agencies. A lack of training opportunities can lead to additional challenges for the use of technology, which often requires extensive training for effective use.

Law Enforcement

SRTB law enforcement agencies reported a range of challenges associated with recruitment, retention, and staffing. For instance, several law enforcement interviewees from border areas felt that their agencies were understaffed, while one municipal police chief from a border area noted that the agency serving the county surrounding his city had few staff, putting limits on their ability to mutually cooperate. Another police chief from a border area added that the low number of staff meant that officers in his agency could not specialize and were forced to be able to address every issue that arises.

Recruiting officers was particularly difficult for SRTB agencies because many applicants are not qualified for the position: "One of the biggest challenges we are facing . . . is the pool of candidates [who] meet the qualifications and pass the background check to be police officers." The underqualified nature of many applicants was often related to the city or town's population size, which provided only a small pool of candidates. Agencies also have difficulty recruiting police candidates who are representative of the population. As one police chief from a border area noted, "One of the biggest challenges we face along the border is a sufficient pool of candidates who qualify to be police officers, especially those who are African American and Latino." A police chief from a border area felt that the desert location of his agency was not very desirable for potential recruits. Law enforcement interviewees suggested that low pay contributed to problems in attracting and retaining officers, which, in one of the border agencies, was being remedied by raising the hourly rate.

In the event that an officer is hired to work for an SRTB law enforcement agency, the officer's tenure in the department was expected to be shorter than in other agencies, for several reasons. Officers might leave to go to larger departments because they offer more-competitive salaries and benefits than SRTB agencies can offer—indeed, Liederbach and Frank, 2003, reports that small and rural agencies offer significantly lower salaries than large agencies. For this reason, command officers often felt like their agencies were more of a "training ground" for officers before they transferred to other organizations. As one officer stated, "We sometimes become a stepping stone for large agencies." The officers who were trained "become low-hanging fruit for nearby city and county agencies to hire them away from us." Another officer stated, "We always have the challenge of keeping the department fully staffed. Inevitably, the larger agencies around us can pay more and have more equipment and provide more training."

Six tribal law enforcement interviewees noted staffing challenges. These included having an insufficient number of staff (mentioned by three interviewees) and training and retention (three interviewees). One police chief shared that his agency had seen 13 different chiefs in the past 20 years. As for possible reasons for low retention, one interviewee pointed out that other agencies could provide better benefits and lure trained staff away. In addition, the same interviewee noted that officer retention was better when officers were hired from within the tribal community because they had a stronger investment in the community and were better accepted by the community.

The isolation of many SRTB agencies means that staff typically need to take on multiple roles. Unlike larger departments with specialized units, officers in SRTB agencies must be more self-reliant and must often serve as generalists. Describing this point, one officer stated, "We don't specialize. I don't have a guy who [handles] sex offenses, homicides, etc. We all have to do it all." Another officer described how it falls to the officers to learn technology:

> In most small agencies and particularly tribal agencies . . . you wear a number of hats. . . . It becomes really challenging to be a jack of all trades with technology. Every purchasing decision you make and every program involves some type of back-end support.

Because of limited staffing levels at SRTB agencies, there can be little time left for training in technology and other areas. As one officer indicated, "Staffing levels make it difficult to provide the right kinds of service and keep up with the rest of the world. Keep pace with technology. Keep pace with mandates." Further augmenting the challenges associated with technology is the lack of IT personnel in SRTB agencies to help with researching, purchasing, acquiring, installing, and maintaining different technologies. As one officer indicated, "I can buy the equipment, but then I have to find someone to pay to set it up. In larger departments, they typically have someone to do that for them."

Because of both funding and staffing deficits, officers often cannot attend conferences and stay informed and trained about technology. As one officer mentioned, "So learning what's out there, obtaining what's out there, and being trained on what's out there is probably the biggest problem that small agencies have." Prior work at RAND found that one of the major needs of law enforcement agencies of any size was improving knowledge of technology and practice within the law enforcement community; the interviews demonstrate this to be true for small and rural agencies specifically (Hollywood, Boon, et al., 2015).

Two police chiefs from tribal areas noted that their agencies did not have dedicated technical staff. One of the chiefs noted that this required him to contract out various technical jobs, such as installation and maintenance of equipment, while the other mentioned that technology maintenance is added to his officers' other duties. In one instance, geography exacerbated perceived staffing challenges—as one chief explained, his agency oversaw three noncontiguous patrol districts that were 80 miles apart. Wakeling et al., 2001, also identifies staffing challenges among tribal agencies and further suggests that these challenges limited tribal agencies' ability to provide round-the-clock services for the entire geographic area they covered.

Staff might also lack skills in other important areas. For example, interviewees from border areas emphasized the shortage of and need for staff with language skills. One police chief stressed that his agency lacked Spanish speakers. Another police chief noted that one-third of the local population in his area are not English speakers. An interviewee from a tribal law enforcement agency explained that, because of financial constraints, the agency does not have a crime analyst who would look at data and trends and that, as a result, the agency was "very reactive rather than proactive."

As a result of these and other challenges, SRTB law enforcement agencies are often understaffed and their officers overworked. Consequently, officers can burn out and leave the department. As one officer stated, "It is difficult to stay fully staffed, so we are dealing with overtime and burnout from people working too much."

Courts

Like their law enforcement counterparts, clerks from rural, small, and tribal courts noted that hiring qualified staffers was difficult. Tribal court clerks said that hiring preferences for tribal members compounded this problem. Staffing levels at one tribal courthouse had declined from five full-time employees to one full-time and three part-time staffers. Another tribal clerk is the only administrator for her court, and she was part time for many years, working full time only when grants funded her position. Related to this issue of understaffed tribal courthouses is the fact that court clerks "wear many hats." One tribal court clerk noted that she also acts as the probation officer, bailiff, court recorder, and jail administrator.

Court respondents also emphasized staff challenges related to the lack of opportunities for training, particularly because the implementation of new technologies places high demand on staff schedules. Training is typically hosted in larger, urban jurisdictions, so, to receive professional technology training, remote courts must send their staff across long distances, which interrupts local staff coverage. However, without adequate training, SRTB court staff might encounter steeper learning curves than their urban counterparts.

Many courts, like law enforcement agencies, have a growing need for reliable interpreters, translators, and bilingual staff. In the words of a court interviewee from a border area, "I think California, and the whole Southwest, Mexican border, there's the need to have bilingual staff and a sufficient number of interpreters."

Some courts reported a lack of local legal service providers, such as lawyers and bonding agencies. One report produced by the Arkansas Access to Justice Commission confirmed the trend toward fewer private-practice lawyers in rural areas in that state. In 2015, the commission and Arkansas' two law schools proposed incentives to encourage new law school graduates to work in the state's rural areas. The interview findings echoed trends in that state, which are relevant to access to justice in rural areas throughout the country. The ABA's work in Wyoming, mentioned above, is another example of efforts to address access to justice in rural areas.

Institutional Corrections

Interviewees from tribal institutional corrections facilities discussed problems in recruiting and retaining sufficient personnel. Tribal facilities rely on the BIA's staffing analysis team to inform personnel decisions. A team of BIA officials who review facility operations and demands to determine personnel needs for each facility execute staffing analysis. After it completes the staffing analysis, the BIA "mandates" the number and types of positions needed to meet agency standards. However, the mechanisms by which the BIA supports or enforces these mandates are unclear. There appears to be a disconnect between what the BIA funds, what the BIA recommends, and what the facility needs. For instance, one administrator was informed after staffing analysis that he needed to double or triple the number of personnel to adequately staff his facility. Yet, even with the recognition that the facility was severely undermanned, the BIA did not provide the resources to increase staffing. Another administrator conceded that these mandates are more appropriately considered recommendations to which funding is loosely tied.

In addition to managing the number of staff, tribal facilities must account for certification and training. During every hour of operation, tribal corrections facilities are required to have on duty at least one corrections officer who has undergone specialized BIA training. This additional training is conducted during six-week sessions in Artesia, New Mexico. One tribal agency mentioned that his facility has not been able to place enough corrections officers into the training session because of a lack of space in the program. Consequently, the facility relies on the subset of its staff who have completed the BIA training to constantly be on duty.

Community Corrections

Remote geography contributes to difficulty with staff retention for community corrections facilities, a problem that many respondents mentioned. If staff are required to work in a central agency office every day, they might not be as likely to take the job or stay in it as long as a job that is closer to home or one that allows them to work out of the home. Agency personnel were considering how to make teleworking effective for officers in order to increase staff retention: "[I] would like to be able to offer staff one day a week for working at home . . . a huge benefit for them, especially with a low salary." Other agencies supported teleworking with the implementation of videoconferencing for staff meetings for officers in remote locations.

Resource constraints also meant that agencies were limited in the number of staff they could hire or the amount of overtime officers could incur. Interviewees frequently reported this as a driver for technology acquisition—to increase efficiency and allow small staff levels to provide sufficient client supervision and support.

Getting IT support can also be a challenge for community corrections agencies. Nearly all of the agencies represented in our interviews were too small to support their own, in-house IT departments, and many rely on the county-level IT departments for IT support. Even some agencies operating at the state level relied on their local counties to provide IT support; although most said that their county IT departments were helpful, they also lamented that their problems or technology issues often received lower priority than those of other departments. Also, county IT staff might not fully understand the unique needs of the community corrections department in order to provide appropriate decisionmaking support. Agencies often rely on state and national association conferences to learn about new technology.

Reliance on IT departments outside of an agency means that a lot of the decisionmaking fell to the chief probation officer or head of the agency, which can be good or bad; the chief probation officer might have a better idea of what technology is right for that office but might not be technology-minded or have the expertise to make good decisions regarding technology. One interviewee reported having made some bad decisions related to technology because of his lack of knowledge: "The process of selecting and acquiring new technology is the chief's responsibility. . . . This has been a very challenging process." Another interviewee said that, "being in a rural area, we learn about stuff after everybody else has it. [The staff would like to] have people come in and teach them what they could be doing differently."

Communications and Information-Technology Management

Many agencies seeking to implement technologies faced challenges in IT management. These challenges included antiquated case-management systems (CMSs) and jail-management systems (JMSs); a lack of standardized electronic data-collection technology across SRTB agencies; and management and analysis software problems caused by poor interface design, needless complexity, or a lack of desired capabilities. All sectors provided evidence of insufficient IT capacity.

Law Enforcement

Many SRTB law enforcement agencies described their technological capacity as being insufficient for carrying out their missions. Some of the technological deficits included a lack of laptop computers and cameras in patrol cars, LPRs, and radios. One officer, for instance, stated that

> the other technological issue we have is that we don't have computers in our police cars. As a result, we are unable to implement community policing projects or CompStat projects. Officers have to take notes in the field and then come back to the station to create that report on a computer. It really isn't cost-effective for us.

Two prior published studies found similar issues—that small and rural agencies are less likely to use, or to see as useful, some technologies, such as GPS and in-car computers (Eastern Kentucky University, 2002; NIJ, 2004).

Many law enforcement personnel from tribal areas described problems with communication technologies, such as radios. One police chief explained that his agency's patrol vehicles did not have radios and that cell phone coverage was not available in his area. Satellite phones could be an option for the agency but were not always available or reliable. As a result, the most-common means of communication in tribal areas was for the officers to ask people to let them use phones in their houses. Another police chief noted that radio communication was possible only in approximately 75 percent of the area for which his agency was responsible. One police chief mentioned that radio communication was problematic in his area because of local geography until a recent $10 million upgrade to the system. Two police chiefs based near the U.S.–Canada border described limitations on their radio communications caused by interference with Canadian communications: "We have radio restrictions because we are along the border. They don't want our radio signals interfering with Canadian police frequencies." One of the chiefs also mentioned that his agency at times received roaming charges because of the proximity of Canadian cell phone towers. Furthermore, because border agencies were in proximity to Canadian law enforcement agencies, radio restrictions inhibited their radio-frequency strength.

Tribal agencies also highlighted problems with inadequate radio systems. As an officer stated, "A big problem for us down here would be communication. We have two portable radios and no radios in the cars." Another officer explained, "We need that capability [radio communication]. We are going to have someone killed because we can't talk to someone quickly." One published study identified the same issue and found that officers address this issue by relying on personal cell phones—more so than do those in larger agencies or urban areas (Foreman,

2013). However, even this solution might not be sufficient because cell phone coverage can be lacking in extremely rural areas.

Interviewees from two border law enforcement agencies discussed record-management software. In one instance, officers could not access the software on their in-car computers; in another, the interviewee lamented that law enforcement agencies did not use the same system, frequently hindering collaboration. Another agency, which was, along with other agencies in its area, in the process of procuring a joint RMS, addressed the same issue.

One police chief based near the U.S.–Mexico border felt that, although border agencies' technological capabilities had improved substantially in past years, organized-crime groups and cartels still possessed far superior equipment. He suggested that requiring manufacturers and providers to work toward common technology standards without proprietary protections would go a long way toward addressing some of the challenges that border (and other) law enforcement agencies face.

Equipment maintenance could also be a problem. One interviewee from an agency serving a seaport described problems in maintaining communication equipment. Because of corrosion, it was very expensive to maintain a fleet of vessels in salt water, and cameras overlooking the port bay were not very useful in inclement weather.

Tribal interviewees' accounts of the technological challenges revealed wide variation across agencies. The technological challenges ranged from cases in which rudimentary equipment (e.g., radios in cars) was not available to cases in which the issue at hand was being able to upgrade or build on existing equipment. Two interviewees shared that they did not face any technological challenges at all. This suggests that agencies' financial capabilities differed substantially, with some agencies struggling to afford relatively basic equipment and others using comparatively expensive technologies.

Courts
A key IT problem that SRTB courts face concerns CMSs, which courts can use to manage case and client information. Many court personnel lamented that their CMSs are antiquated or outright dysfunctional. According to these respondents, CMS components that should work together are internally inconsistent or incompatible. For example, the financial system is not linked to the CMS, requiring manual intervention. Some of these systems can be difficult to reconfigure and adapt to statutory and policy changes. Several clerks at tribal courts complained that their CMSs were subpar, reflecting sentiments similar to other small, rural, and border courts, while two tribal courts lack CMSs entirely.

Tribal court respondents noted other problems related to technology, including the inability to access Westlaw, lack of court recording software and video teleconferencing technology, and even phones without conference-call capabilities. In addition, some of the tribal clerks mentioned poor Internet connections, spotty cell phone coverage, and lack of landlines. Communication with residents is therefore a problem. Two tribes also mentioned that their websites are out of date.

Institutional Corrections
Technology planning and support challenges in institutional corrections facilities were reported to affect both current operations and future technology adoption. Many agencies lack sufficient bandwidth to run systems, including RMSs, video visitation, telemedicine, and remote arraignment. These problems lead to increased transportation and contracting costs, as well

as added stress to staff and inmates because services must be conducted off-site. Slow Internet speeds also hamper opportunities to offer inexpensive Internet-based services, such as job training and legal research, to inmates.

However, even when new technology is planned for implementation, ongoing support from vendors and the cost of maintenance can both be perceived as risks. Several respondents reported that vendors understate costs or fail to provide long-term support. Others reported having limited vendor options because of their remote locations or dealing with local vendors that were not as experienced or competent as a larger vendor might be.

Many agencies face long-term problems because they lack IT management expertise. Although some facilities found success taking on projects themselves or hired entrepreneurial and knowledgeable staff to research and implement new technology, other facilities reported that "folks [who] have absolutely no background in technology are making decisions." Another jail administrator explained the problem of technology planning in SRTB areas:

> It's difficult for small counties because we don't have the IT leadership or the engineering leadership to design the systems so that 14 years from now we aren't in the same boat. So we've been buying pieces off the Internet, wherever we can find things. If I were to give a grade to our system, I would say [a] D. But I don't think I'm unique in that area. A lot of jails that are built, there's no vision for the future. It just doesn't feel like we have the leadership or knowledge in that area to make good decisions.

Institutional corrections facilities reported mixed success with their JMSs and RMSs. Their representatives often reported JMS platforms not to be user friendly, not to function correctly, or to require connections to existing systems that are not possible with many older jails' current infrastructures. For example, some tasks that might be logged in real time must instead be logged later on because of the complexity of some types of data entry. Many facilities use data systems based on law enforcement, which were not designed with corrections tasks in mind. Although common tasks should be streamlined, custom reports—designed to address needs better than the preloaded options—require programming expertise that many agencies do not have. Another issue is that data systems might not provide data in a usable format: "One of the main problems I have with the RMS is that I can't get raw data out. . . . It's aggregated, and I can't break down numbers into smaller categories." Redundancy and complexity were also common complaints. For example, using one prominent RMS requires users to "type in the same inmate information on six different screens." Another interviewee reported that, with his agency's system, one must "go three screens deep to get a booking number."

Tribal jail administrators seemed aware of available software and hardware that would vastly improve operations but that they did not currently use—typically because of lack of resources to obtain and maintain the technology. In addition to wanting better information-management and -sharing systems, administrators mentioned needing to increase the quality and number of cameras in their facilities, migrate log-keeping and notes to an electronic system, and improve Internet coverage and staff access to computers. However, as a few administrators explained, "ultimately budget and funding are considered the root causes of challenges"; "technology is not the priority, it's a privilege. After personnel, staff, payroll, everything you need for inmates, including clothing and food, leftover money goes to wraparound services, then after that, which is usually nothing, comes technology."

Community Corrections

Community corrections respondents commented extensively on IT use and limitations, often highlighting their struggles with technologies that are considered standard in larger or urban agencies. The interviews made clear that agencies need mobile devices of some sort to improve their ability to work while not in the office and that most agencies were considering, testing, or using a variety of devices. The range of options and considerations noted (e.g., cost, connectivity, security), however, suggests that decisions about which technology is best for each agency varies significantly.

Poor cell phone coverage limited some agencies' ability to use remote monitoring devices. One agency reported piloting remote breath testing using a cell phone to take a picture of the offender during the breath test but experienced problems with data coverage. Another agency head was hesitant to implement new technology because "there are places we don't have cell service, so some technology wouldn't work anyways."

For some, this was addressed through radios in officers' cars, "but even those [the radios] have dead spots." Those in especially mountainous areas had significant trouble with cell phone reception, with one interviewee pointing out that the spotty cell phone coverage in his area was "a safety issue."

Some interviewees reported skipping smartphones altogether in favor of tablets, which allowed easier navigation and viewing of forms while in the field. Most interviewees reported having at least some laptops or tablets available for officers; some reported that officers found the laptops too bulky, while others said that tablets were not secure enough to allow officers to access files while in the field.

Many agencies reported limited Wi-Fi access for officers while in the field or could not afford a data plan for all officers' cell phones or remote devices: "We don't have the money to do the data plans." Officers thus rely on available Wi-Fi signals to be able to connect to the office and receive emails. In remote areas, however, Wi-Fi can be hard to locate and can provide unreliable connections, rendering their mobile devices less helpful. Several community corrections respondents reported insufficient bandwidth to fully utilize existing technology acquisitions (e.g., video visitation technologies and RMSs linked to external databases) or consider new technology solutions.

Regardless of whether agencies allowed remote data access, all agencies with officers spending time in the field reported that remote access was or would be helpful in improving officer efficiency in the field. For example, "maybe if they [officers] have nonarresting contact with someone, if they could access the records during that kind of interaction, that would be very useful to them."

One agency also suggested that the ability to set automated reminders for individual clients about meetings, court hearings, or treatment would help keep clients on track: "It's human nature to . . . put your head in the sand when you don't want to do something. So you try to take away all excuses why [the probationer] forgot the court hearing." This could help reduce failure-to-appear charges and help clients successfully complete supervision. Further, many agencies are responsible for PSIs and have officers who have to spend time in the local jail meeting with offenders and filling out paperwork. For these officers, remote access to data files, the ability to complete electronic forms on laptops or tablets, and collecting electronic signatures would be especially time-saving.

Agencies commonly reported a desire to go paperless—although most also indicated that, because of required connections with systems beyond their control—such as with state oversight agencies or interactions with paper-based court systems—they had not been able to go fully or mostly paperless. Interviewees reported that electronic file access would make it easier to collect needed signatures on tablets or smartphones while in the field and send completed forms into the office without requiring the officer to take a paper version into the office or into court. Electronic files also would support more teleworking.

Data-Sharing and Interoperability

Related to IT challenges are problems with data-sharing and communication in the field, including problems with interoperability. Many respondents across agency types reported that not being able to access the data they needed to do their jobs well impeded their ability to carry out their duties.

Law Enforcement

As discussed above, many SRTB law enforcement agencies identified communication in the field as a primary challenge for their departments. One of the frequently discussed challenges was the need for interoperable radio communication with other departments and agencies. As one officer stated,

> We have zero interoperability. . . . Think about a spontaneous event that involves people in real time needing information, needing to communicate. . . . This is one of those problems that is just waiting for an incident to happen like 9/11 to illustrate how these two people wearing uniforms couldn't exchange critical information spontaneously when they needed to and someone got hurt or killed. This is a major technological issue.

Poor interoperability with other agencies was described as being the consequence of antiquated radio technology and weak radio-frequency strength, which were often related to funding and the physical geography of the jurisdiction. Addressing this point, one officer stated,

> One problem is our radio communications system. . . . Our radio breaks down at least once a week. This makes it difficult to communicate with our local police officers, the [residents], park rangers, and the department of public safety. We spend a lot of money trying to fix it.

Data-sharing issues in law enforcement go beyond problems of interoperability. A law enforcement interviewee pointed out that many officers in border (along with small and rural) agencies do not have the security clearances required for close cooperation with federal agencies, such as the FBI, although most of the officers are eligible. He suggested that the reason for this might be local police chiefs' unfamiliarity with the required procedures of obtaining clearance or the lack of working relationships with the federal agencies in question. A related challenge is working and communicating with law enforcement across the border where there might be issues with trust across agencies.

One police chief from an agency near the U.S.–Mexico border noted high turnover and corruption among Mexican police as an obstacle to effective collaboration, forcing his agency to keep some distance from his Mexican counterparts: "We work with them but keep them at arm's reach because of the corruption. We have to be careful about sharing information."

Courts

Several court respondents noted that a lack of data-sharing infrastructure prevents judges and qualified staff from electronically accessing important databases. For example, judges in small courts might lack authentication controls needed to access case records. As a result, they must resort to manual queries or rely on the clerk of court to conduct electronic queries for them. A similar problem reportedly arises when small courts lack the scale necessary to afford flexible software license agreements, often asking multiple staff members to share a single computer's license.

Some courts, owing to their small number of full-time staff, cannot support any operations outside standard business hours. This limitation reportedly delays officers conducting urgent duties, such as carrying out emergency child custody orders over the weekend, and prevents clients from accessing court information outside business hours. One respondent described the situation in her state:

> [In this state,] if a cop runs across a child abuse or neglect situation during nonbusiness hours, they can't remove child without a court order. If it's 2:00 a.m., they have to be able to call a judge and have [the] petition reviewed. At one point, cops would have to drive to [the] judge's house for a face-to-face meeting. Often a waste of time. A lot with technology can be done with using fillable documents that can be prepared and sent over a mobile device—as long as you can get an electronic signature.

A lack of web infrastructure at some courts makes it difficult for the public to access public court records. A tribal respondent explained,

> We have no access to state criminal justice databases, so judges often don't know of outstanding bench warrants or parallel cases in other jurisdictions. [Our reservation] is close to other reservations, but we do not have good coordination mechanisms with those reservations. So when people get in trouble in one reservation, they can just hop to the next one and avoid prosecution.

Respondents attributed this problem to insufficient resources to build and maintain a court website or public-facing web tools on their existing websites. This limited web functionality also restricts the court's ability to conduct important community outreach.

Many electronic systems lack interoperability with external partners, making it difficult to share information with other agencies, such as local law enforcement, probation, and parole. As a result, such agencies might not be informed when a person in custody is eligible for early release. As one respondent explained,

> We have multiple systems, only one of those I would categorize as a modern or current case-management system. The rest of them date back to the '70s and '80s, and they're not robust, they don't support electronic filing, they don't support document imaging system, document-management systems, so we're in bad shape, and we're working to address that.

In a similar vein, one tribal clerk mentioned a lack of communication between reservations, despite geographic proximity. He explained that people easily skip from one reservation to another to avoid prosecution. These reservations all have individually customized CMSs, and he notes that a shared system would help avoid these problems.

One court referred to the inefficiencies that traditional dispute-resolution procedures bring about. Compared with innovative practices that online competitors (e.g., LegalZoom) use, traditional offline procedures are slow, redundant, and expensive, which delays access to justice. One respondent explained how the court was not getting as many civil dispute cases because it lacks an online dispute-resolution capability:

> The last five years of civil cases, we're seeing confirmed nationally a trend we suspected was gaining speed, which is that civil cases are leaving the system at an accelerated rate. We think this is due to the widening gap between what people are willing to pay and complexities they're willing to endure and what is actually required at the court. People are leaving the court for civil disputes. Some of this business is going to commercial online dispute resolution; seeing small claims and divorces going that way too. Ten years ago, it was only the high-end complex civil cases that went to private arbitration, but now [we are] seeing it much more broadly. Starting to think that courts need to provide similar online dispute-resolution capability, to retain that kind of business.

One factor affecting court staff and judge information access was whether the court is a part of a unified (i.e., statewide) or decentralized court system. Small courts in decentralized states tended to more frequently voice difficulties with data-sharing and interagency coordination. The upshot is that these decentralized courts admit to more autonomy while some of their unified-court counterparts reported frustration about whether state-level policymaking authorities hear their unique needs or suggestions. As one respondent noted,

> The state doesn't have a clue what we do at this level. [State officials] need to come down here and work in one of these offices for a week. I feel like this local level should train the state, instead of the state training us.

An NCSC report (NCSC, 2010) and a RAND report on technology needs for courts (Hollywood, Woods, et al., 2015) both concluded that electronic access to court records was a significant need among courts of any size; the JIC interview findings suggest that this is especially true for small and rural courts.

The large flows of people and goods through border areas and the nature of work of border criminal justice agencies frequently require border agencies to collaborate with other entities, both within the country and internationally. This often leads to operational challenges involving data-sharing and interoperability. For example, court interviewees from border areas described how immigration-related offenses might require increased collaboration with federal law enforcement agencies:

> If a [defendant] is an illegal alien and detained here for a state offense . . . once [that defendant is] done with probation, they would be deported. But that is another law enforcement entity that does that. We only take care of the defendant until they get placed in probation and immediately deported; as for the length of time from [probation to deportation], that is unknown to me. That would be customs protection [sic] whenever they have time or when there is, I can't really speak for them but is taken care of by customs border protection [sic]. As to a defendant who is also an illegal alien also would be sentenced to time to serve, the individual would carry out sentence at the [state] penitentiary. That would be under another [state] department of criminal justice. That would be out of our hands.

Because populations often move fluidly across the borders in these areas, court respondents noted that sometimes border courts must engage with courts and agencies that operate across international borders. For example, border courts might be required to enforce orders originating across the border and ensure that legal orders are recognized across jurisdictions. These issues would be relevant across states as well:

> We have issues involving families and children that can cross back and forth fairly ubiquitously. Our judges have to work with authorities in Mexico, when we've got kids who are wards of the court, divorce issues, where parties move fluidly back and forth. . . . We do occasionally have cases where, "How do you enforce orders from Mexico?" and vice versa, but more focused on apprehension and prosecution of criminal cases and, to the extent where we have issues involving dependents and wards of the courts and divorce, custody, trying to achieve family objectives in the cross-border environment.

Despite problems in collaboration, court respondents were generally positive about their relationships with officials across the border, noting that these cases tended to be relatively rare and are not unique to border courts:

> Our contacts across the border are fairly limited. It hasn't come to mind where we've had notable issues communicating across the border. Challenges communicating across agencies aren't particular to being a border community. Our judges may need to contact resources in Mexico, and we've seemed to be able to do that. We've had judicial officers go to Mexico and vice versa, to share ideas and share things. It's never come up as a major problem.

Institutional Corrections

Representatives from institutional corrections agencies also expressed a need for better interoperability and external information-sharing capabilities. One respondent noted that systems purchased from multiple IT vendors often do not work together as promised:

> Our vendors will tell us they will work together and then, when the rubber hits the road, they don't. Because we are not IT people, often we don't know exactly what we're looking for, so when we get it, we're disappointed. I think that a lot of our smaller agencies have many different vendors. . . . Well, they are not actually going to work together. So we spent a lot of money for nothing.

Some facilities have access to statewide data systems, but this was not the norm. Data, including records from court agencies, law enforcement, and neighboring jail and prison facilities, would help jail staff better identify inmate needs and risks. A respondent defined the potential for such an arrangement:

> One thing that we have is this vision about us working together with neighboring counties to identify inmate discipline [disciplinary action taken against inmates for behavioral issues]. Track that from agency to agency. There are statewide agencies that do a little bit but not enough to make classification decisions.

Data-sharing limitations seemed to have a particularly acute effect on tribal agencies, which just received access to federal databases in 2015. Most, however, still lack access to local offender data, which can present safety issues. Some tribal respondents noted the lack of access to complete criminal or incarceration records of their inmates. One administrator relayed a story about how the facility received an inmate on public intoxication charges who had, unbeknownst to booking and leadership, been convicted of murder decades earlier in another jurisdiction. Furthermore, the brother of the victim in that murder case was housed in the facility at the same time. If not for a corrections officer recognizing the inmate and alerting his superiors, the consequences might have been severe. Tribes' current inability to access criminal and incarceration records from other tribes, as well as from nontribal agencies, such as local and state law enforcement, creates major safety and security risks for inmates and corrections officers alike.

Border corrections agencies reported having to work with a different set of criminal justice agencies from those working with most nonborder agencies, which exacerbated shortcomings in data-sharing and interoperability with other agencies, including local law enforcement, and federal agencies, including U.S. Customs and Border Protection. In some instances, agencies do not share information on criminal and incarceration histories, including previous offenses and risk-assessment data. Several police chiefs, including two who explicitly stressed bad communication with federal agencies, echoed the lack of information-sharing:

> We do not have interoperable radio communication with National Park Service or [U.S. Customs and Border Protection Air and Marine Operations]. This is a major issue because we cannot communicate in real time with these agencies. This creates extra work and risk for us. That is one of those problems that is just waiting for an incident to happen like 9/11 to illustrate how two uniformed individuals couldn't exchange critical information spontaneously and got hurt or killed.

Related to risk assessments is the fact that corrections agencies have to address unique residency issues and determine how that affects risk: "Because many probationers are from Mexico . . . during pretrial, should we let them reside in Mexico?" Another interviewee added, "Sometimes there are residents [who] live in Mexico but are U.S. citizens. If they live in a different country, there is no way to supervise them. Our officers can't cross into Mexico." Absconders, then, are a bigger challenge for agencies in border areas than for other agencies. Tribal facilities also face challenges caused by a lack of access to other facilities' data, as discussed in the example above about the victim's brother in the same facility as the convicted perpetrator, with this fact being discovered not through data exchange but because a corrections officer recognized the former.

Community Corrections

Nearly all interviewees noted limited data access and data-sharing with other agencies in their states as common and significant challenges. Those in agencies allowing (or wanting to allow) remote file access cited challenges, including lack of access to a CMS that is web-based and thus accessible through any web browser, or a lack of apps created by the software vendor that would allow remote access to the data system through a smartphone or tablet. Some agencies—especially those using a required statewide CMS—reported that the system could not be accessed remotely and that officers had to use computers in the office through a virtual private network link to a central server: "It is extremely difficult to access this program in the field. You basically have to be on site to access it." Others reported that even using a virtual private network could be difficult: "It is cumbersome."

Among agencies using mobile devices that allowed data access, interviewees were mixed on whether access to confidential information should be allowed. Some reported their own concerns about the security of mobile devices: Technology "is very important, but security and confidentiality of information are important, so you have to be careful about that too." Others reported that their IT departments would not allow remote data access because of security: "You have to get special permission to access the CMS while in the field."

Very few reported working in states where officers could readily access offender records regarding prior or current supervision in other counties. Some reported access to limited information and the need to call an officer in another county for details—especially in those states where every county in the state can choose its own CMS. Others reported significant frustrations with state systems that kept counties in individual silos, preventing an easy flow of information between different state offices: "[Other agencies] generally fax over information, which our agency has to enter into our own system. We can't share information electronically," and "[We] do not have connectivity to other counties through our CMS technology."

Some reported that their states were moving toward statewide CMSs that would allow easy data-sharing across jurisdictions: "The new system will help with communication between probation departments because they'll be able to see other counties' cases," and "[Sharing data across counties using a statewide system is] not difficult at all. . . . We can tap into those things pretty easily." Others reported that data-sharing across counties was unlikely to be easy or to happen any time soon: "I wish the state would recognize the benefit of having a sole system," and "The new [state] system does not give us any information that we need."

Despite these challenges, data accessibility and sharing were on the minds of most interviewees: "There is a lot of detriment to not knowing what's going on with a person in other places." Limits on data-sharing among different agencies about active clients mean that an officer might not know whether a client is on probation in another county or has other restrictions from another county and might not be able to readily get information on a client's performance on probation in another area. The need to share data was especially relevant for those officers conducting PSIs; in some agencies, those officers had access to different systems with more-complete offender histories. Jackson et al., 2015, which identified the need for community corrections agencies to improve "fragmented" systems of information-sharing and improve data extraction and use from existing systems, confirms JIC interviewees' reports.

Community corrections agencies are required to work closely with courts and corrections agencies in order to operate effectively. Whether they need access to offenders in jail to conduct PSIs or coordinate services postrelease, need to work with the courts to determine the best technologies for supervision, or need to file PSIs or report on offender success on supervision to the courts, difficulties often arise when agency cooperation is necessary. Agencies reported a mixed bag of experiences working with these other agencies: "[My biggest challenge is] communication with law enforcement and [the] court." Agencies clearly have a need to make interactions with these other agency types as easy as possible.

Beyond being able to work remotely on laptops or tablets from the courthouse, interviewees identified a range of areas in which court technology could help their work, including an electronic docket that would allow better notification of hearing dates and times, minimizing officer wait time in court. They also noted electronic filing of paperwork and electronic signatures as areas of need. One interviewee reported receiving handwritten court information, which, based on officer time in court, might not get entered into probation's CMS until the next day. Another reported needing a better way to share information in the other direction: "We would like to e-file our presentence reports to decrease the amount of paper reporting we have to do," and "We need something easier to use because the judge and county attorney all [sic] want copies of our documents."

Other agencies reported relying on law enforcement for equipment, such as laptops or portable breath testers. Some agencies sent clients to the police station for required alcohol breath testing, and others received input from law enforcement on technologies, such as portable breath testers, that law enforcement use more regularly: "We rely on law enforcement."

Thus, working with other agencies in the criminal justice system is difficult but required: "When you are out and networking with the schools, law enforcement, you have that information to exchange, and that is how the criminal justice system works—the greater information you have, the more reliable the system is."

Crime-Specific Challenges

SRTB agencies often have to deal more frequently with particular types of crime, as interviewees from law enforcement agencies, courts, and institutional corrections facilities highlighted.

Law Enforcement

Several SRTB agencies identified various crime-specific challenges within their communities. Officers reported problems related to alcohol and drug use, domestic violence, and drug and human trafficking. As one officer stated, "We're battling a never-ending issue with prescription drugs, heroin, and amphetamines, just like any other agency out there."

Furthermore, for some border agencies, drug and human trafficking were major challenges. Illustrating this issue, one officer stated, "We have information that drug and human traffickers use the . . . mountains as a route, which is pretty much confirmed by border patrol." These drug trafficking and human trafficking challenges were not isolated to land. As one officer noted, "We have a substantial challenge with human and drug smuggling in the maritime . . . area. As security has increased on the landside border, there has been an increase in the maritime area for smuggling people and drugs." These comments echo findings from the literature on human trafficking and the struggles of enforcement agencies to adequately address such problems and to identify and assist victims (Owens et al., 2014).

Three tribal police chiefs cited high rates of poverty, substance use, and violence in their respective communities as a challenge. Although this issue is not specific to tribal agencies and indeed some small and rural interviewees reported it as well, it is worth reiterating that tribal communities fare worse on a range of health and social outcomes than their nontribal counterparts (Beals et al., 2005).

Other crime-related issues that respondents noted include the prevalence of casinos in some SRTB areas, which some interviewees felt created and attracted a significant amount of crime.

Three police chiefs operating near the U.S.–Mexico border highlighted money, drugs, and human trafficking and prostitution as substantial problems in their areas. One interviewee mentioned drug trafficking as an issue but explained that the problem typically originated in the contiguous 48 states rather than Canada. A few tribal interviewees offered examples of needs and challenges that might have pertained to very specific contexts, such as trespassing and illegal trash dumping in archeological sites in the area that the agency served, and crime surrounding popular areas, such as casinos and a National Hot Rod Association–certified race track.

Courts

A border court respondent mentioned that increased attention was required for specific crimes, such as drug trafficking, and stressed the lack of adequate funding across all stages of the justice system. This is another issue that could affect courts throughout the United States but might be exacerbated in border areas, where concerns about such issues as immigration or trafficking in illicit goods are heightened.

For example, one participant in a border area noted that law enforcement and prosecutorial funds had increased for apprehension and prosecution of drug offenses, which had caused an increase in workload for court staff. As a consequence, other cases could lose priority and attention from the court:

> For example, the [district attorney] has grant monies coming in to prosecute drug offenses in border areas, and we're prohibited from applying for similar grant funding. When you fund one part of the justice system and not other agencies that are part of the system, you can create dislocations and inefficiencies that put us right in the middle. No guarantee that the state will step in and fill the gap. They're getting additional human resources to pursue prosecutions and prepare cases and bring them to court, but whether it's the court or probation system, the people who have to deal with the aftermath of those prosecutions, that can impact other cases in criminal or even civil areas, where you have to divert resources to what's being filed by prosecutors. [This] can affect the defense bar because you need people to defend those cases.

Institutional Corrections

Institutional corrections agencies located in border regions expressed an increased need for contraband- and drug-detection technology, but modern technology solutions, such as the full-body scanners that the Transportation Security Administration employs at airports, are too expensive for most SRTB agencies to consider. This is a growing concern because the nature of drug and contraband trafficking into the United States is constantly evolving.

Infrastructure Challenges

Limitations in facility and technological infrastructure were common in SRTB agencies, particularly courts and institutional corrections facilities. Many facilities were reported to be old and in need of repairs and renovations in order to be suitable for technology upgrades. Other facilities were so antiquated that some technology acquisition would make sense only when a new facility is built. Facility shortcomings often created security concerns for employees, with some respondents reporting unreliable utilities, even including electricity and Internet connectivity. In our sample, these issues were mentioned with more frequency among the court and jail interviews; other needs were more dominant in the law enforcement and community corrections interviews (among the latter, this issue was not mentioned at all).

Courts

Many court personnel commented on the poor condition of their facilities. Several respondents reported that they lack basic security measures, such as security staff, barriers between defendants and judges, and screening technologies, such as metal detectors.

Some courts, particularly tribal courts, faced challenges with the structural adequacy of their buildings. One tribal court was operating out of a triple-wide mobile home that lacks any space for a jury and leaks during the rainy season. As a result, one respondent noted, "None of the eight [tribal areas] has held a jury trial for last 20 years. We don't have room for a jury; we have a living room good for watching Sunday football, but really not for a court." Another tribal respondent described a "traveling court," which is housed in a conference room with exercise machines in the back, while another described leaks in the courtroom facility.

Some counties have very limited infrastructure for safety and security. As one respondent noted,

> For counties, there is no push and no one holding them accountable. As court administrator, I can talk until blue in the face to say, "please, we want more security, badges," and make sure everyone is safe, but we can't fund it ourselves. It's a big deal and very frustrating.

Further, basic utilities, such as electricity, can be unreliable in some SRTB courts, making it extremely difficult to operate at times. As one respondent stated, "The power grid is 'dirty.' There are lots of surges that need to be protected against for the tech systems." Many of these facilities also lack adequate security. As noted earlier, some courts, especially those in small or rural areas, lack basic technological infrastructure, such as broadband Internet and cell phone coverage. Without these basic services, respondents report, adopting other technologies that rely on them, such as GPS monitoring, is difficult.

Institutional Corrections

Interviewees reported managing a range of facilities; some had newly built facilities, while others were working in facilities that were more than 50 years old, and the age of the facility seemed to determine much about the level of technology present. Many older facilities present challenges for staff based on construction, layout, and space limitations.

Many tribal corrections administrators indicated that they oversee facilities that are ill suited for their roles. One administrator explained that the facility that the team currently inhabits was originally built as a rehabilitation center in the 1970s. The facility does not have a central command and control area, adequate space for visitation, or proper housing for female inmates. Another administrator explained that the ceilings in that facility were so low that it is impossible to install cameras in the pods, where they would not be at risk of tampering by inmates. One administrator mentioned that, even though the facility was recently built, the layout impeded supervision of inmates and management of the facility. There, the control area is on one end of the facility—not in the center—and the size of the facility creates a long travel time from one wing to another.

Recent changes in response to the Prison Rape Elimination Act (Pub. L. 108-79, 2003) (PREA), solitary confinement, and segregation needs—including changes implemented for medical reasons, safety reasons, or disciplinary issues—have heightened the need for changes in technology and related infrastructure in institutional corrections facilities. For example, most respondents mentioned that they had just upgraded, were in the process of upgrading, or needed to upgrade their cameras because of new PREA rules. Even if a respondent did not mention PREA explicitly, it is a driving force for the focus on camera improvement. In older facilities, poor sightlines, narrow corridors, isolated areas, low ceilings, and outdated wiring that are more common to such facilities complicate challenges related to coming into compliance with PREA and improving surveillance in general.

Renovations for new equipment, such as wiring or Wi-Fi repeaters for high-speed data and Internet access within the facility, are often prohibitively expensive and disruptive, especially for facilities operating at or near capacity. Some interviewees reported not planning on implementing new technology until a new facility was built—which could be several years into the future: "Everything right now hinges on whether we build a new facility or not." Others reported having to implement technology piecemeal, over time, in order to be able to afford it, such as updating wiring one year and installing cameras the following year, or installing half the cameras in one year and the other half the following year, because budgets do not support a full upgrade at one time.

Resistance to Technological Change

For many SRTB agencies, the rarity of discretionary funding available for technology acquisitions, along with the lack of technological infrastructure and training opportunities, contributes to the slow adoption of potential technology solutions, even for known needs. These themes were noted among court and corrections respondents in our interviews.

Courts

Many court respondents described resistance to implementing technologies for routine court procedures. Several court respondents noted that the heavily used practice of manually filing forms can cause confusion and delay among the growing number of pro se (i.e., self-represented) litigants. Similarly, manual record management, reporting, and payout increases court caseloads because of the large time investment required of staff. A respondent noted,

> The time constraints of when someone fresh to legal issues comes in and wants to file pleadings—it takes a lot of time to administratively guide these people through the process. We're seeing more and more self-help and hands-on in-person. . . . Monitoring all that, staying on top of that, is a huge responsibility, very time-consuming, and it has to be accurate.

As a result of manual filing, many people miss court deadlines and dates—a problem that an automated reminder system could avoid. In addition, for some courts, paper records are digitized by scanning documents individually, rendering digitization projects costly and time-consuming. In addition, manual court reporting presents a bottleneck for some courts, given the shortage of court reporters in some geographic areas.

NCSC's guides to rural court technologies suggest that online systems can help address some court tasks, such as citations, tickets, and fines, and reduce the need for some people to travel to a court in person. Such a system could alleviate some of the bottlenecks that interviewees from small and rural courts reported.

A few court respondents highlighted the tension between judges and administrators surrounding the use of particular procedures or technologies. They described some courts as slow to adopt innovations because of a culture of distrust of technology and the fact that some staffers are content with the current technologies. Sometimes, a previous bad experience with technology leads to further resistance. For example, one respondent noted,

> The problem with videoconferencing historically is that it's not always done well. If not done well, then it's a bad experience. Judges have low tolerance for bad experiences with technologies during live court hearings. [They] tend not to go back if they have a bad experience.

Some jurisdictions want to adopt new technologies or procedures but are hampered by political opposition. One respondent, for instance, favored the use of treatment services as an alternative to incarceration but did not receive the necessary political support to implement this practice. Effective technology implementation, one expert emphasized, requires not just technical expertise but also strong institutional leadership. As one respondent stated,

> Another barrier or challenge is people's natural adversity to change. That's why you need someone with a vision, able to bring people along with them. Having some strong governance group in a particular court or area that would help figure out how to clearly identify your goals and how technology could help you and how to implement efficiently and get around roadblocks.

Federal and state organizations have worked to overcome this issue through education on technology and other innovative problem-solving solutions, providing seminars to court professionals to improve justice in rural areas (Mahoney, Carlson, and Baehler, 2006). With more-widespread knowledge on available technology and solutions that improve rural courts, these strategies could gain more-widespread acceptance.

Community Corrections

Among community corrections agencies, many were focused less on new and emerging technologies or very sophisticated technologies, such as kiosks for remote check-ins or remote GPS, alcohol, or drug monitoring, and more on improvements to basic technologies. For some, the focus on basics, such as phone systems, Internet access, and computers, arose out of need—many agencies seemed far behind in terms of the sort of technology standards that are common in larger agencies.

For others, the focus on basics arose because they did not see the need for more-sophisticated technology. For example, one interviewee talked about a kiosk check-in system: "We considered it, but we aren't that busy to justify it with the number of clients"; another reported, "We don't use [a kiosk] because so few people are coming into the office at one time that we don't really need it." Another common theme was that agencies are "getting by with what we have." Even if they know about technology, many are not sure that it is appropriate, that they could afford it, or that a mobile technology would work in a rural area with limited connectivity.

Interviewees had very mixed responses regarding smartphones. Some agencies issued them to all officers; others only to supervisors or officers with high-risk clients. Others were in the process of piloting smartphones, and still others either did not feel that they were necessary or did not have interest from staff or the budget to support an upgrade to smartphones: "It's difficult to get older staff to move into the tech age," and "[older officers] just didn't get it." The primary concerns with providing smartphones, among those who did not already use them, were the security of files accessed remotely, the uncertainty regarding how necessary smartphones actually were, and—with a tight budget—the concern about paying for a data plan in addition to the cellular service.

Legal and Policy Barriers

Interviewees from law enforcement agencies and courts in particular noted challenges created by external policies and mandates. Examples from law enforcement and corrections include mandates to increase camera utilization without sufficient funding for the implementing agencies. Other policies forbid technology options with potential benefits, exemplified by document e-filing, which is prohibited in many courts despite enthusiasm about the potential reduction in administrative costs from respondents. Court respondents in particular commented on this issue. Tribal agencies face some legal barriers because of the way their agencies are administered.

Law Enforcement

Interviewees reported some issues stemming from policy developments. Two police chiefs mentioned that their states' public records act required personal information to be scrubbed when a member of the public requests a camera recording. This created a potentially labor-intensive task for officers. An interviewee from a different state echoed this concern about costly requests for information, although not in relation to any concrete policy measure. Also in relation to policy surrounding the use of video, one chief explained that their state was a two-party consent state, which put limits on officers' ability to make and use recordings.

A law enforcement interviewee based in a Southwest border area felt that, because his agency dealt with immigration issues, political fallout stemming from the national policy debate on the issue and recent policy developments affected the agency:

> We have the political fallout that comes with putting the police in the middle of immigration issues. For instance, the [state law requiring law enforcement officers to determine someone's immigration status during lawful stops in cases of reasonable suspicion that the person is an undocumented migrant] created antipolice culture, it made it very difficult for us to deal with our Hispanic community that is from Mexico. We have seen an increase in antipolice rhetoric and demonstrations in light of the immigration issues. The Hispanic population would be reluctant to report a crime for fear of deportation.

Tribal law enforcement agencies noted a range of challenges related to governance and tribal sovereignty. For instance, one tribal police chief explained that there was no existing tribal domestic abuse law his agency could enforce. Instead, prosecutable cases were handled as sexual assaults or batteries. One chief mentioned that the tribal community was guarded about its sovereignty and that its members did not want to share much information with outsiders and did not necessarily trust the police department. As a result, the chief felt that intratribe communication between her agency and the tribe was inhibited. Another tribal police chief said that the tribe that his agency serves was reluctant to accept limits on its powers and that, as a consequence, the chief's primary challenge was informing the tribe of its legal parameters ("because they are sovereign, [members of the tribe] feel like they can do whatever they want").

Another chief noted difficulties with government-to-government cooperation because his agency was established relatively recently. He found agreements, such as memoranda of understanding (MOUs), between his agency and nontribal partners difficult to establish, resulting in challenges, such as getting the local jail to accept the agency's arrests and convincing the prosecutor to take up the agency's cases. In addition, the agency did not have a dispatch agreement in place with local and state police "because [those agencies] are not used to entering into an agreement with a sovereign nation." Instead, they used cell phones to communicate. This lack of documented policies and procedures and the challenges it can present has also been identified in prior research on tribal law enforcement agencies (Wakeling et al., 2001).

Courts

Court respondents cited legal and policy barriers as significant obstacles in the implementation of technologies and best practices. One respondent explained that many activities, such as the signing of court documents, legally require a physical presence at courthouse. E-filing by lawyers, for example, is reportedly not authorized in that jurisdiction, and implementation of this practice would require a statutory change at the state level. Another respondent noted that some CMSs can be difficult to reconfigure and adapt to statutory and policy changes.

One of the interviewed experts said that state laws requiring a court in every county regardless of its caseload facilitated the economic struggles of rural courts. This legal imperative highlights a perceived tension between justice access and financial sustainability.

Another challenge, one that is unique to tribal courts, has to do with issues of concurrent jurisdictions. These problems manifest differently among the tribes. One clerk explained that her tribe, which has a service area but not a reservation, is in a Pub. L. 83-280 state, meaning that the state has jurisdiction over criminal prohibitory issues but not enforcement of orders. Law enforcement in this tribe is virtually nonexistent. Another clerk noted that, because her reservation encompasses four different tribes across two states, "coordination and jurisdictional issues . . . are constantly coming up." Another aspect that these clerks mentioned is tribal courts' inability to access the state court's criminal database. Courts therefore cannot run background checks, which means that decisions about child custody and other issues are made without full information.

Another distinctive challenge that one clerk in a tribal court described and related to the unique position of tribal courts in the judicial system was that the tribe had five different codes. It was unclear what the law actually was, and the clerk had to gather disparate information and rewrite portions of the law in order to try to codify that reservation's laws.

Conclusion

Taken together, the SRTB interviews revealed a broad range of operational challenges that clustered around a few central themes. These tended to involve difficulties capitalizing on the human, financial, informational, and technological resources needed to operate safely, effectively, and efficiently in environments that are geographically, economically, and politically limited or unpredictable. These challenges tend to have negative impacts on their downstream investment options and their communities' access to justice.

Funding shortfalls were universally mentioned and colored many of the other challenges that our 12 communities of practice face. Respondents throughout the spectrum of participating agencies reported that small budgets imperiled recruiting, retention, or training. The rarity of discretionary funding for technology acquisitions also contributed to slow adoption of potential technology solutions to known needs. These agencies suffer another disadvantage: They usually cannot take advantage of the economies of scale for technology acquisition that larger metropolitan-area agencies often do. As a result, these agencies often struggle to take on technology projects requiring large capital investments.

Rural and border agencies reported geographic isolation, which led to specific challenges ranging from getting jurors and defendants to courthouses on trial dates because of long travel times and a lack of transportation resources in the community, to border law enforcement agencies combating drug cartels with superior technological capabilities. Geographic isolation also creates staffing problems because of transportation times for police officers responding to calls, community corrections and court officers conducting community supervision, and jail staff transporting inmates for external health services and court appearances.

Agencies reported facing greater obstacles in IT management, including antiquated CMSs and JMSs, and less technology across SRTB law enforcement agencies than their urban counterpart agencies. Management and analysis software problems caused by poor interface design, needless complexity, or a lack of desired capabilities were also reported widely. All sectors further provided evidence of insufficient IT capacity among personnel.

Facility and other physical infrastructure limitations were common in courts and institutional corrections. Many facilities were reported to be old and in need of repairs and renovations in order to be suitable for technology upgrades. Other facilities were so antiquated that some technology acquisitions would make sense only when planning for a new facility. The facility shortcomings often created security concerns for employees. Some agencies reported unreliable utilities, including even electricity and Internet connectivity.

Difficulties with information-sharing and interoperability across agencies were also widely reported. In border areas, this problem extended to tactical operations, including communications across systems to administrative needs, including court, criminal, corrections, immigration, and medical and treatment record integration. Many respondents reported that their inability to access the data they needed to do their jobs well impeded their ability to carry out their duties.

Language barriers between practitioners and persons in the criminal justice system were common across sectors, especially in border areas. Finally, courts and corrections agency representatives indicated a lack of transportation, job training, substance-use treatment, mental health treatment, and medical services for clients.

Many operational challenges that tribal agencies face are similar or identical to challenges that interviewees from nontribal small, rural, or border agencies describe. In addition, these agencies also face some unique challenges, particularly related to governance and tribal sovereignty.

In the next phase of JIC's research agenda, described in Chapter Five, a select panel of experts met to discuss their perceptions of SRTB needs as our interview respondents described. They evaluated these themes based not just on their prevalence but also on their degree of importance and how amenable they are to change. Consideration of each of these criteria will be necessary to develop a useful discussion about the technological solutions and innovations and their application to SRTB agencies.

CHAPTER FIVE

The Justice Innovation Center Advisory Panel: Identifying Science- and Technology-Related Needs to Address Pressing Issues

Through a literature review, interviews, and attendance at conferences, JIC has highlighted key issues that SRTB agencies face, as reported in the previous chapter. We now address the extremely important question of what might be done to address those issues. JIC's advisory panel, made up of experts and practitioners representing the different agency and sector types, was tasked with addressing this question, identifying and prioritizing a set of science- and technology-related *needs*. We identified the panelists in a variety of ways, including recognized experts in the field who work with professional associations, professional association membership, practitioners who have collaborated with RAND or NIJ in the past, and particularly knowledgeable interviewees. We purposefully selected members who would represent each of the 12 communities of practice and had complimentary experiences (see Table 1.1 in Chapter One).

Here, we define a need as a requirement for innovation to address a problem (or opportunity) that an SRTB agency faces. The prospective innovation was to be related in some way to science and technology but might include technological development, development of guidance and educational materials, development of training, development and dissemination of new business practices, and development and dissemination of model policies or even legislation. The parties asked to carry out the innovation might include NIJ but could also include other DOJ agencies, other federal agencies, commercial technology developers, academic researchers, and practitioner associations.

To generate and assess the needs, the advisory panel followed an approach developed through a set of prior RAND studies on criminal justice technology needs (Goodison, Davis, and Jackson, 2015; Hollywood, Boon, et al., 2015; Hollywood, Woods, et al., 2015; Jackson et al., 2015; Silberglitt et al., 2015). Prior to the meeting, panelists were sent read-ahead material summarizing the top issues that SRTB agencies face, culled from our various information-collection efforts. We then asked attendees to identify any additional major challenges that SRTB agencies face—that had not been specifically identified in the interviews—as well as technology-related ideas for addressing the issues that should be brought up in the meeting.

The JIC advisory panel convened on December 7 and 8, 2015, at RAND's Arlington, Virginia, office. It devoted day 1 of the meeting first to reviewing the key issues that SRTB agencies face and initial ideas for solutions. Participants spent the rest of the day in four breakout groups (one each on law enforcement, courts, institutional corrections, and community corrections) to foster discussion specific to the needs of each agency type.

Day 2 of the meeting focused on prioritizing needs for each agency type. We asked each participant to complete a survey specific to that participant's agency type. The survey asked participants to prioritize—or rank order—each need identified in the discussion groups by importance for their specific agency type. Participants also had the opportunity to write comments as to why they rated a need the way they did. After we aggregated participants' initial rankings, JIC researchers developed the overall prioritization (ranking) for each need, resulting in a list, for each agency type, of needs ordered from most to least important or relevant.

JIC used a formula to group the ranked needs into one of three tiers: tier 1 (high priority), tier 2 (medium priority), and tier 3 (lowest priority). Several days after the workshop, panelists had the opportunity to take a second online survey. This questionnaire showed participants the tier for each need, which was based on their initial rankings, along with the comments that panelists had made for each need, and offered participants the opportunity to vote needs into higher or lower tiers (higher or lower priorities). We then adjusted the rankings by taking into account each up or down vote, to produce a final ranking of needs in three priority tiers.

In looking at the issues that the interviewees raised across all agency types (as discussed in Chapter Four), we identified four top-level issue areas. Specific challenges can belong to two or more of these overarching issue areas. The issue areas are as follows:

- shortfalls related to *IT*, including interoperability, infrastructure, and difficulties upgrading to new systems
- shortfalls related to *agency operations*, including policies, governance, business processes, training, recruiting, and retention
- shortfalls specifically created by the *geographic dispersion* of many SRTB agencies, including the challenges to having to cover large areas, travel long distances regularly, and having comparatively few services and resources locally
- shortfalls created by a *lack of funding and other resources* that can be especially acute for SRTB agencies.

We then captured the major reported shortfalls within each of these themes, resulting in a list of key issues for each breakout group, which JIC provided to the panelists in their read-ahead packets prior to the meeting. Table 5.1 summarizes the major shortfalls that were covered in each issue area by agency type. For the complete lists provided to the panel, please see Appendix D.

Needs That the Advisory Panel Identified

Through discussions in the breakout groups, participants identified a large number of needs for each agency type. JIC generated and prioritized a total of 239 needs, as follows:

- law enforcement: 66 needs
- courts: 75 needs
- institutional corrections: 52 needs
- community corrections: 46 needs.

Figure 5.1 shows the proportion of needs in each technology issue area—IT, operations, geography, and resourcing. Approximately one-third of all needs focused on IT and one-third on operations. The remaining third was split between funding and geographic issues. Within each agency type, however, different patterns emerged (see Figure 5.1). Law enforcement and

Table 5.1
Summary of Major Reported Shortfalls Covered in Panel Read-Ahead Materials

Agency Type	IT	Agency Operations	Geographic	Funding and Resourcing
Law enforcement	• Interoperability • Special interoperability issues for border agents • Infrastructure • Impacts of large areas and long distances • Difficulties in adopting new technologies	• Lack of specialization • Difficulties in recruiting and retention • Difficulties in relationships with the community • Issues specifically for border agencies • Issues specifically for tribal agencies	• Lack of key local resources • Impacts of large areas and long distances	• General • Technology funding • Federal funding • Special issues for border agencies
Court	• Interoperability shortfalls • Infrastructure shortfalls • Difficulties of adopting technology • Language barriers	• Process innovation • Access to justice	• Distance challenges • Shortfalls in local capabilities in rural areas	• Revenues • Equipment • Staffing
Institutional corrections	• Interoperability • Infrastructure • JMSs • Qualified vendors	• Mental health and substance-use treatment • Staffing	• Transportation	• Budget • Facilities • Addressing health needs
Community corrections	• Interoperability • Infrastructure • IT support • Remote data access	• Electronic files • Substance-use testing • Office management	• Transportation	• Budget • Service provision

institutional corrections needs focused heavily on operations; court needs focused heavily on IT issues; and community corrections needs focused heavily on resourcing issues.

We also considered the breakdown of needs that participants classified as tier 1 (high priority) need in the ranking exercise; Figure 5.2 shows these results. Here, we do see substantial differences between agency types, with operational issues dominating top institutional corrections needs. Eight tier 1 institutional corrections needs concerned a single operational issue—providing mental health and other treatment services to inmates. Geographic issues played a larger role for courts, and resourcing played a larger role for law enforcement and community corrections agencies.

Appendix E provides the full list of tier 1 needs, sorted first by issue area and then by agency type. Within categories, needs are shown in priority order. Each need includes an operational problem and a corresponding specific need to develop and field a potential solution.

Key Themes from the Panel Discussions

Two general points of interest cut across all the breakout groups' discussions.

The first is that the majority of issues and needs discussed were not SRTB-specific; they apply to larger agencies as well. The results were broadly consistent with needs identified

Figure 5.1
Needs, by Issue Area

Agency type	Geography	Information technology	Operations	Resourcing
Law enforcement	6	30	47	17
Court	7	43	36	15
Institutional corrections	12	31	44	13
Community corrections	11	26	28	35

Percentage of needs identified

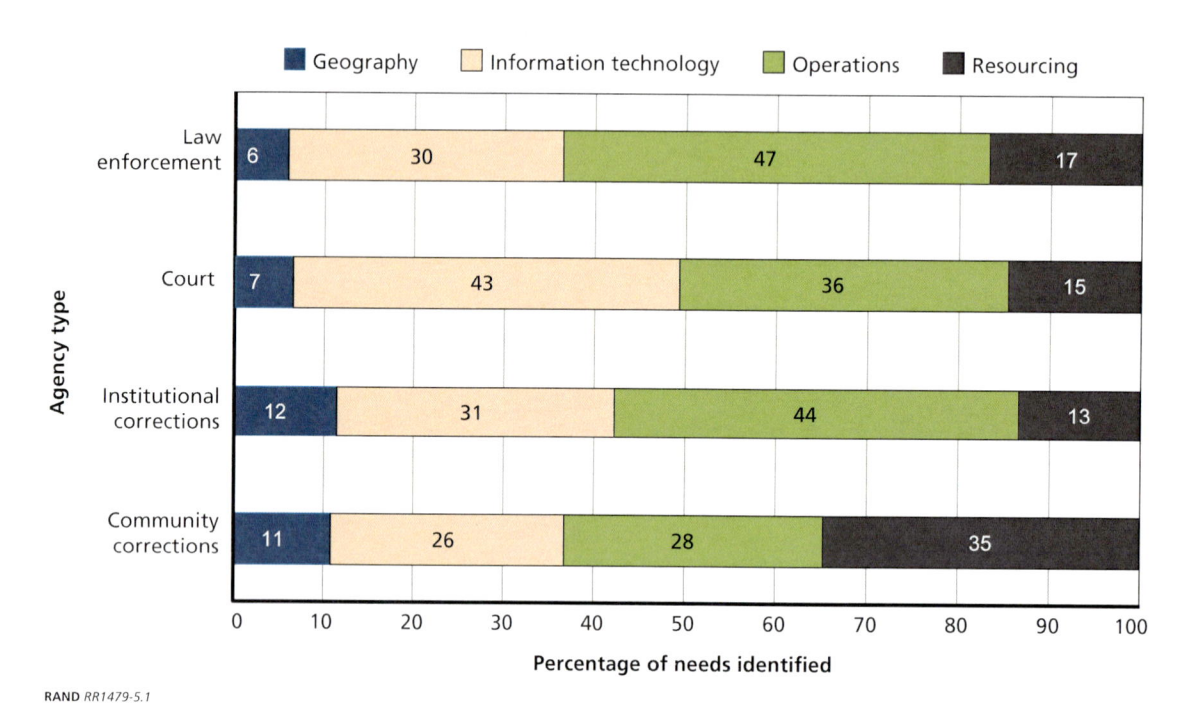

Figure 5.2
Tier 1 Needs, by Issue Area

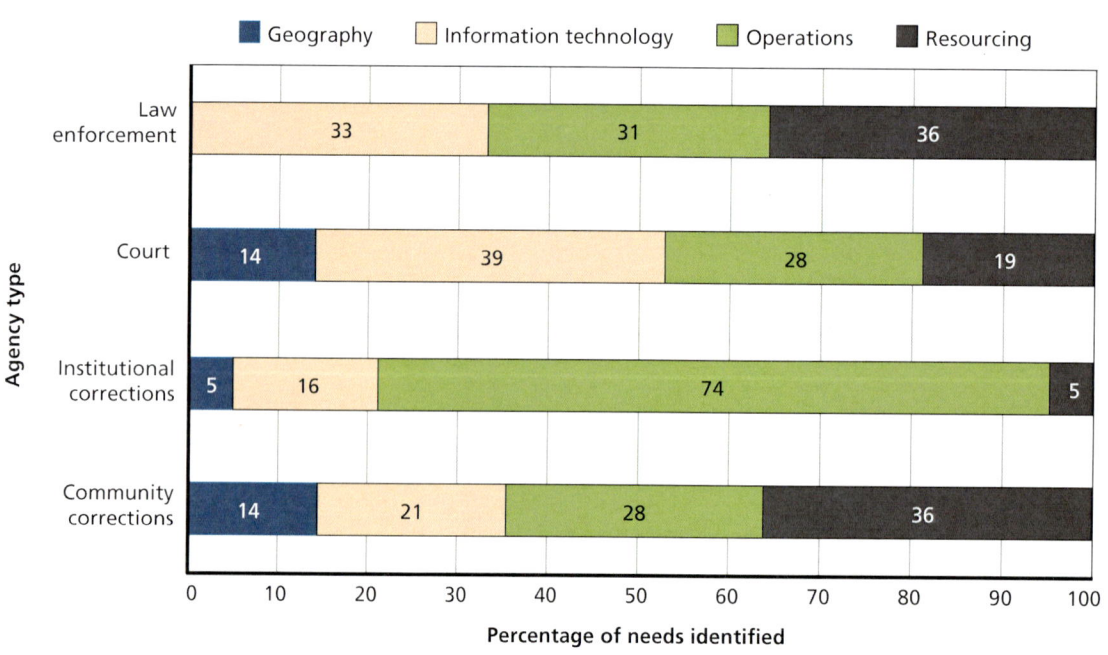

Agency type	Geography	Information technology	Operations	Resourcing
Law enforcement		33	31	36
Court	14	39	28	19
Institutional corrections	5	16	74	5
Community corrections	14	21	28	36

Percentage of needs identified

through RAND's Priority Criminal Justice Needs Initiative (Hollywood, Boon, et al., 2015). All breakout groups had top-ranking needs related to improving the sharing and use of information, as well as needs for providing educational support to improve agencies' knowledge of technology and how to acquire and use it effectively. Notable exceptions had to do with the needs related to geographic challenges (agencies have to cover long distances; people with key skills and service providers are not typically local) and resourcing challenges (smaller populations meant that SRTB agencies were extremely resource challenged; small size further reduced agencies' capabilities to acquire technologies and seek grants and other funding). Tribal and border agencies also reported some sector-specific challenges. Under "Crosscutting Themes" later in this section, we discuss specific themes and needs that cut across multiple communities of practice.

The second point of interest is that panelists from all four agency types reported being challenged by some sort of major societal or cultural change. For law enforcement, it was the need to improve relationships with the broader community. For courts, it was the need to support an influx of litigants representing themselves (addressing litigants who did not speak English was a related concern). For institutional corrections, it was the need to provide greatly improved mental health and other treatment services to inmates. Finally, for community corrections, it was the need to focus efforts on rehabilitation and positive behavioral change rather than continued punishment.

It should also be noted that the themes and needs do not map exactly onto the ones identified in the interviews because the experts were allowed to submit their own needs both prior to the meeting and during the sessions.

Law Enforcement Themes
Improving Relationships with the Community
The general need to improve relationships and build trust with the community was a very strong theme in the panelists' discussions. The top-rated law enforcement need was to improve agencies' abilities to work with the media. Another need was to disseminate strategies on how to increase agencies' presence in their communities. Technically oriented needs included developing social media applications that allow members of the community to provide feedback and increasing the use of community feedback surveys.

Improving Information-Sharing for Law Enforcement and Leveraging Common Standards
Specific needs called for assessing the feasibility of state and national standards for sharing state- and federal-level data with agencies and developing corresponding patches for legacy systems to incorporate those standards. There was also a call to develop models for becoming compliant with the National Incident-Based Reporting System (NIBRS) standard for reporting crime data.

Addressing Information-Technology Management Shortfalls
The top-ranking need here was to develop tools to help small agencies better determine their IT needs and match those needs with specific systems and vendors. Other needs specifically addressed IT system management. These included assessing mechanisms that enable vendors to administer agency systems remotely, assessing remote IT training opportunities, and developing low-cost solutions for fostering internal IT management and administration capabilities.

Improving Communication Infrastructure
Participants highly ranked a variety of needs here. The first was to consider mechanisms for replacing older analog equipment with newer digital equipment in order to improve effective

range. The second was to assess setting up a web-based marketplace for agencies to sell older but still effective communication equipment to SRTB agencies needing it (similar to eBay). The third was to work with the Federal Communications Commission (FCC) to ensure the consistency and availability of needed spectrum; some participants discussed problems with getting strongly needed frequencies assigned from the FCC in certain areas that (1) had been approved elsewhere and (2) did not appear to have any interference problems.

Getting a Better Understanding of Body-Worn Cameras
Participants called for additional information for SRTB agencies on the purchase, use, governance, and costs of BWCs.

Addressing Responses over Large Geographic Areas
Panelists saw the problem of responding over long distances as a high priority. The top-ranking need that emerged was the development of a model MOU for mutual aid to permit neighboring agencies to respond under specified conditions; note that this idea was somewhat controversial with panel members.

Improving Resourcing
Participants recommended assisting agencies in building grant-development capabilities, both in general and specifically through TA centers that would help SRTB agencies with funding applications. Outside of grants, participants asked for assistance in identifying ways of financing equipment purchases in general. Participants specifically suggested pursuing the use of state procurement programs to help agencies save money.

Court Themes
Addressing the Surge in Pro Se Litigants
Participants saw addressing the influx of pro se litigants, many of whom had no experience with court processes and proceedings, as a major challenge. Specific needs here included developing a Turbo Tax–like app that would help litigants complete legal forms, as well as developing guides, brochures, and videos that would explain court procedures to the public.

Somewhat related to this theme was a need to expand SRTB courts' capabilities to work with litigants who did not speak English. The top-rated need here was to explore ways to have interpreters certified for other fields (e.g., health care) rapidly certified to be translators for courts.

Improving Security and Resiliency at SRTB Courts
Several top-ranked needs applied to improving security and resiliency in SRTB courts. Three needs concerned physical security. These included providing continuity-of-operations planning tools, providing emergency response training, and evaluating courthouses' abilities to respond to emergencies.

Improving Information-Technology Infrastructure for Courts in General
There was a strong desire to build up courts' IT infrastructure in general. As noted above, the top-ranked need for courts was to lobby local governments to build up their Internet infrastructure. There was also a high-rated need to assess whether courts could buy systems to leverage new First Responder Network Authority (FirstNet) capabilities.

To assist courts directly, participants called for assessments of the use of satellite communications in areas not well served by cellular service. Similarly, there was a call to assess using general-purpose, low-cost communication systems (such as voice-over–Internet protocol services) as potential replacements for purchasing customized communication systems. There was a related call to assess whether courts could use low-cost communication tools—notably, social media tools—to communicate with remote populations. More broadly, participants called for developing "reliable videoconferencing that is simple and works every time."

Improving Information-Sharing for Courts
Participants stated a need to develop model policies and a planning process to determine what information will be shared and how, especially with the public, in developing new CMSs.

Improving Technology Training
There was one top-rated need to increase the use of online training courses for court staff, leveraging existing examples (Utah was identified as one potential model). Another called for creating and funding model technology or mentoring courts specifically for SRTB agencies. Participants noted that, although model technology courts exist, they are usually models for large jurisdictions.

Addressing Funding Shortfalls
Finally, there was a desire for increased technology-related funding of courts. One top-rated need here included funding for technology training, while a similar need called for funding of IT-related TA. The second requested that grants cover "broad and innovative uses" of new systems, rather than require funded systems that would be used for only a single purchase. The aforementioned call for local governments to build up their Internet infrastructures in general can also be seen as fitting in this category, as was the call to fund technology model or mentoring courts.

Institutional Corrections Themes
Improving the Provision of Mental Health and Other Services to Inmates
Providing care services for inmates, especially mental health care, was one of the principal themes of the institutional corrections breakout group's discussion. Participants discussed a variety of needs in support of this theme. The top-rated need from the group was for jails and academies to provide deescalation training. A related need was for jails to set up, train, and use specialized critical-incident or crisis-intervention teams to respond to inmates in crisis. There was also a call to provide agencies with a mental health assessment tool for incoming inmates that can be used to support decisions regarding care and placement.

Several needs related to getting external support for mental health care for inmates. The first was to help agencies develop good relationships with external mental health agencies. The second was a call for state-level mandates for agencies to work with jails to improve care for mentally ill people. The third was for assistance in creating partnerships with external agencies to help place inmates with mental health problems. A final need was for agencies to create handbooks of services and resources (including housing, employment, and physical and mental health) that could educate inmates about available services prior to their release. To provide mental health services over long distances, participants called for materials on best practices for the implementation and use of telepsychiatry. To help gain momentum for providing care services, participants said, agencies need to analyze inmate data to assess the impact of the mentally ill on their operations and on the criminal justice system in general.

Developing Corrections Personnel

The need to improve the capabilities of both agencies and individuals (as well as individual retention) was another top theme from the institutional corrections breakout group. As noted, deescalation training was the most requested type of training, with needs related to both initial deescalation training (the top-rated need from the breakout group) and ongoing deescalation and self-defense training. The second-rated need was a call for agency heads to better develop line staff (the front-line staff, such as corrections officers and patrol officers) and actively involve them in the workings of institutions (by involving them in committees, for example). The third-rated need was a call for increased academy training specifically on corrections as opposed to law enforcement.

From an agency-wide perspective, there was a call for agencies to employ American Correctional Association (ACA) standards. Panelists felt that, even if agencies were not seeking formal certification, standards could still be useful in informing agencies about best practices.

Improving Information-Sharing, Especially Via Improving Jail-Management Systems

Needs here began with a call for a checklist of mandatory elements that should be in a JMS. Specific elements that panelists requested included generating data for state reports, generating data for federal reports, and providing for visual display of hot spots within a facility. This was followed by a need for developing a program for certifying commercial providers as experts in corrections IT systems, who could provide systems that were genuinely focused on institutional corrections operations. Finally, there was a call for funding IT experts who could provide TA on corrections systems.

Addressing Funding Shortfalls

Panelists noted that agencies have to devote substantial time and resources to seeking funding and conducting procurement. The specific tier 1 need in response requested that federal grant regulations be modified so agencies could hire part-time employees to administer grants.

Community Corrections Themes

Refocusing on Rehabilitation and Positive Behavioral Change

Participants emphasized repeatedly that the most important takeaway should be that community corrections has a broad-based need to transition to focusing on rehabilitation and positive behavioral change, not continued punishment, as its main purpose. This starts with a change in terminology; participants noted that community corrections officers work with clients, not offenders.

Corresponding needs here included, first, a call for the community to redefine the role of officers as a combination of law enforcement and social work, eliminating items outside of that scope. There was a similar call for officers to be able to connect clients with people who could provide "role-modeling behavior" in their communities. From a technical perspective, participants asked that performance information from monitoring technologies (e.g., GPS tracking, substance-use tests) be used to provide positive reinforcement, not just to punish lapses. Participants also asked for new risk-assessment instruments that would not just assess a client's risk of reoffending but specifically link the client to evidence-based treatment programs that would maximize that client's chance of success. More broadly, participants asked for research that would assess what types of monitoring of and data collection from clients would be most useful, in order to limit the burdens on officers, agencies, and clients.

It is also important to note that many of the top-ranking needs categorized under different themes fell in the context of refocusing community corrections. For example, discussions of ways to make better use of remote technologies focused on how those technologies could provide rehabilitative services to clients.

Improving Information-Sharing

Participants focused information-sharing needs on providing the equivalent of a law enforcement mobile data terminal for community corrections officers. The baseline and top-rated need in this area was to develop a tablet or phablet that would include CMS functions and integration, along with access to other key databases and other task-specific software. (A second tier 1 need also called for giving connected laptops, tablets, or smartphones to corrections officers.)

The second need focused on the back end of providing information to community corrections officers, i.e., on building statewide or regional platforms to conduct agency, state, regional, and federal information-sharing. The biggest feature of this platform was an alerting system that would update corrections officers when high-interest events happened (notably, rearrests). Participants also described the need to create common MOUs concerning necessary policies.

Two needs addressed IT support for workflows. The first was to migrate from legacy paper processes to digital workflows. The second was to provide support for dictation software, so officers could dictate their reports.

Finally, there was a top-rated need to develop and field access-control systems that would allow officers remote access to the information they need in the field while maintaining network security.

Addressing Geographic Dispersion

The four top-rated needs from the community corrections breakout group all focused on ways to overcome geographic challenges. These were all in the context of providing rehabilitative services to clients.

Two needs addressed leveraging videoconferencing, with the top need focusing on the use of videoconferencing for officer–client visits and the other on videoconferencing for mental health and other service providers' sessions. There are also requirements for verified security and privacy. Note that participants emphasized that initial meetings with clients needed to be in-person meetings but that later meetings could be conducted remotely.

A third need called for building relationships with law enforcement so that community corrections and law enforcement officers could reinforce each other's efforts in remote areas. The specific example was for law enforcement to provide breath testing of clients using their existing equipment.

The final need in this area was for agencies to get catalogs of treatment facilities and programs available in their wider region. The catalogs were to include information on program capacity, costs, and effectiveness.

Addressing Funding Shortfalls

Finally, a top-rated need related to resourcing challenges was to provide agencies with support in identifying new grant opportunities and writing corresponding grant proposals. A second related need was to help agencies better assess the feasibility and costs of technology upgrades to be funded through grants.

A third need applied to a different area of resourcing shortfalls. This need called for agencies to develop staffing models that permit the use of support staff (i.e., nonsworn personnel) to support clients' nontherapeutic needs.

Crosscutting Themes

Although needs tended to be tailored to their community of practice, some needs from different groups were close enough to highlight themes cutting across the criminal justice community in SRTB settings.

Improving Information-Sharing Between Agency and Other Governmental Systems

All breakout groups had some top needs calling for improved information-sharing between an agency's systems and information repositories at the regional, state, and federal levels. Needs covered both reporting to these repositories and receiving information from them.

For *reporting*, expert panelists from the law enforcement group called for models to adopt the NIBRS crime data–reporting standard. The institutional corrections group called for JMSs to support state- and federal-level reporting consistently. The courts group more broadly called for planning as to what information should be shared and with whom—specifically, what information would be shared with the public.

For *receiving information*, the law enforcement group wanted to assess the use of standards and implementation of patches for getting access to repository information. The community corrections group similarly wanted access to repository data, especially data on high-interest events that would support alerts to community corrections officers in the field. This group also covered access control and network security.

Assisting Agencies with the Procurement and Management of Information-Technology Systems

All groups had needs to assist agencies with procuring and running IT systems. Law enforcement and courts both had top-rated needs to assess remote training for IT, with courts adding calls for technology mentorship courts and TA funding more broadly, and law enforcement adding calls for a low-cost way to improve internal IT management capabilities in general. The law enforcement group also had a need to help agencies determine their IT needs and match them to specific commercial providers. The institutional corrections group called for training, certifying, and funding IT vendors and consultants.

Using Videoconferencing to Overcome Distance Barriers

The courts, institutional corrections, and community corrections groups all had needs related to assessing and using videoconferencing tools (including commercial tools) for both internal uses (hearings, client meetings) and external uses (treatment sessions).

Assisting Agencies with Obtaining Grants

The law enforcement and community corrections groups both called for activities that would educate agencies about available funding opportunities and then provide TA to help agencies write grant applications. The institutional corrections group also noted having to spend a great deal of time seeking and applying for funding.

Using Nonstandard Personnel

Three breakout groups had needs on assessing the use of nonstandard (i.e., nonsworn) personnel to carry out certain functions at lower cost. For law enforcement, it was remote IT system administrators; for institutional corrections, it was part-time grant administrators; and for community corrections, it was nonsworn staff to handle clients' nontherapeutic needs.

Conclusion

The themes and needs identified here, and especially their priority rankings, set the stage for JIC's subsequent work. The needs will be used to identify field studies that will have the greatest impact on high-priority needs in order to ensure that the center's work has utility for SRTB agencies.

CHAPTER SIX

Conclusions

Technology makes significant contributions to the effectiveness, efficiency, and safety of the criminal justice system. Work to develop new technologies—and to find new ways of using existing technologies—can improve the efforts of law enforcement, the courts, and corrections agencies in many ways. However, the development and application of technology in these sectors can be challenging. For a variety of reasons, this challenge is felt most acutely in SRTB areas. Ensuring public safety and residents' access to justice is often difficult in these areas: Resources are often limited, and the geographic landscape can present significant challenges. At the same time, small agencies and those in rural, tribal, and border settings are less likely to use technologies considered standard in urban areas.

JIC's work and the findings presented in this volume are designed to contribute to an understanding of and address barriers to the use of technology in SRTB areas in order to improve public safety and access to justice.

JIC's efforts have considered the role of technology in SRTB criminal justice agencies through a variety of methods, including a literature review, in-depth interviews with nearly 150 practitioners and topical experts, and focused discussions with an advisory panel of experts and practitioners. This information-gathering work highlighted the limited amount of existing research on the technology in JIC's contexts of interest—SRTB areas.

JIC's work thus far has included gathering additional information on the challenges that SRTB agencies face and prioritizing the needs of criminal justice agencies in SRTB contexts. JIC staff also began identifying relevant technology solutions that might address the highest-priority needs and began planning for ways to assess those technology solutions as implemented in real-world situations; these pilot evaluations were undertaken in JIC's second year.

The main activities of JIC covered in this report focus on four main research goals that are the foundation of JIC's efforts in this area: (1) identifying the unique operational needs of SRTB justice systems; (2) identifying potential, innovative, technology-based solutions to meet those needs; (3) conducting trials to evaluate technology solutions on criminal justice policy and practice; and (4) disseminating the results of JIC's work. Findings from research goals 1 and 2 were the focus of the current report.

Understanding Technology in SRTB Contexts

JIC was developed with the understanding that technology can make significant contributions to the effectiveness, efficiency, and safety of the criminal justice system and that work to develop new technologies—and to find new ways of using existing technologies—can improve the efforts of law enforcement, the courts, and corrections agencies in many ways. However, JIC's work during its first year confirmed that the development and application of technology in these sectors can be challenging—especially in SRTB areas.

JIC's in-depth literature review revealed significant gaps in areas of interest to the center. Most of the existing research on technology in criminal justice agencies focuses on larger agencies and those in more-urban settings. In addition, the bulk of existing research considers technology use in law enforcement agencies; significantly less attention has been paid to technology in courts and corrections agencies. Even less work has been done on tribal agencies or those located in border areas. The JIC literature review also found limited evidence for the effectiveness or utility of technology; few rigorous evaluations of technology have been conducted with SRTB agencies. The limited evidence available leaves SRTB agencies without guidance on selecting clear and defensible technology priorities.

The JIC literature review described the general challenges in these underresearched and underresourced settings, identifying some differences across sectors (SRTB) and justice agency type or community of practice (law enforcement, courts, and corrections) that warranted careful consideration. However, the small body of literature relevant to JIC's work also identified relatively similar challenges across agency types. Prior work suggested that most criminal justice agencies saw a need to focus on improving "fragmented" systems of information-sharing among siloed agencies and improve data extraction and usability for a range of users. Improved training for handling offenders with mental health problems was also common across communities of practice.

The discussion of the SRTB context for criminal justice agencies highlighted the dearth of research that has been done in these areas. Although it was limited, the literature review provided a foundation to help understand the ways in which technologies are already being used, the outputs and outcomes of their use, and the ways in which they could be used to address unmet needs. The review suggested that more attention be paid to the needs of SRTB agencies and informed the development of questions for the subsequent expert and practitioner interviews and the advisory panel discussions.

Gathering Input on Operational Needs from the Field: Interviews

To prioritize the acquisition of technologies for use in criminal justice agencies, and in light of the limited existing research on the topic, JIC researchers first had to gather additional information on the challenges that SRTB agencies face—challenges that might be addressed through technology solutions. To do that, JIC staff reached out to practitioners across the country to get firsthand information on their challenges and hear stories of their technology successes that might be shared with other agencies facing similar issues. From these in-depth interviews—we spoke with nearly 150 sheriffs, judges, court staff, jail administrators, probation office directors, and others in 38 states—JIC was able to pull together a picture of how technology is being used, how it could be used, and how to implement it in SRTB criminal justice agencies. In the process, JIC also learned about attempts to use technology that did not quite work but still provided important lessons on using technology in a criminal justice organization.

As with the literature review, we found a range of needs relevant to each community of practice and each sector, but there was also significant overlap in some of the most-pressing challenges of which practitioners spoke in the interviews. These challenges often compromised the agencies' abilities to provide access to justice for their communities in environments that are geographically, economically, and politically limited or unpredictable.

Funding shortfalls were universal and were the basis for many of the other challenges that respondents across sectors and agency types described. Small budgets affected agencies' staff recruiting, retention, and training. Budgetary restraints were reported to make technology acquisition difficult and slow, especially for technology solutions requiring large capital investments. Moreover, many small and rural agencies reported struggling to meet requirements from new policies or mandates because of limited resources. For example, new camera requirements in many jurisdictions for both police and corrections agencies can strain existing resources and facilities.

Small and isolated agencies reported limited ability to harness economies of scale, limiting negotiating power with potential vendors. Some agencies joined buying collaboratives to take advantage of economies of scale or to reduce their costs and put more-expensive technologies in reach.

Challenges related to information-sharing and interoperability across agencies also were widely reported—confirming the findings from the literature review. Many respondents reported that not being able to access the data they needed to do their jobs well impeded their ability to carry out their duties. All sectors provided evidence of insufficient IT capacity among personnel, further contributing to difficulties in modernizing or upgrading existing systems. Many interviewees also complained about subpar RMSs and CMSs that did not meet their reporting and data-management needs.

The interviews identified a set of needs that SRTB agencies face that went beyond the information in existing published literature and provided new insights on the focus of JIC. We used the information gathered via the interviews to develop the agenda and discussion points for a JIC advisory panel meeting.

Gathering Input on Operational Needs from the Field: Advisory Panel

JIC's interviews, conferences attended, and literature reviews highlighted key issues that SRTB agencies face. To dive deeper into understanding the needs of SRTB agencies and identify technology solutions, JIC convened a group of experts to prioritize the needs. The team classified prioritized needs into tiers 1, 2, and 3 (high, medium, and low priorities, respectively), and, as with interview results, there were some clear overlaps in priority needs across communities of practice. For example, from different perspectives, all agency types identified a need to be better equipped to address mental health concerns among the people they serve, whether that involves training law enforcement or corrections officers to address people in crisis or better reaching people in small and rural areas with needed mental health services via video technology. Agencies also prioritized IT expertise as a significant need; many smaller agencies do not have in-house IT experts. Finally, panelists identified interoperability on a variety of levels as lacking; agencies want to share data with other agencies and jurisdictions, communicate with other agencies and first responders, and obtain better data-management systems. As found with the interviews, panelists identified geographic landscapes as barriers to providing services in rural areas and cited resource limitations as hindering progress on technology innovation.

Identifying the highest-priority needs for each community of practice will allow JIC researchers to identify pilot research projects that will have the most relevance and impact in the field.

Synthesis of Lessons from the Justice Innovation Center's Initial Work

JIC's first year provided important information about challenges that SRTB agencies face, including gaps in existing technologies and in agencies' ability to implement technologies effectively because of such issues as funding shortfalls, slow acquisition of technologies, and constrained data-sharing capabilities. The center's work also identified several high-priority needs for criminal justice agencies in SRTB contexts, including better IT support, improved interoperability, and better capabilities for addressing mental health concerns. Table 6.1 summarizes the key findings from JIC's first year.

JIC's first year of research helped to fill in some gaps in understanding the technology challenges and needs that SRTB agencies face. One important lesson that comes from looking across the findings from the first year is that, although SRTB agencies face many common problems, each agency type also faces some issues that are prominent within that type of agency, such as the need for law enforcement agencies to improve relationships with communities or the need for courts to address the surge in pro se litigants. This means that technology solutions must not only be capable of addressing the broad challenges that SRTB agencies face (e.g., the need to cover long distances, lack of funding)—many of which are shared by agencies of all sizes—but must also be capable of addressing the specific needs of each agency type and of individual agencies.

Another important lesson from the first year is that better data are needed concerning the effectiveness of technologies to address the challenges that SRTB agencies face. JIC's literature review found only a few studies that examine the effectiveness or utility of technology in general, and very few rigorous evaluations of technology have been conducted with SRTB agencies. SRTB agencies need better guidance on how to select clear and defensible technology priorities.

JIC's subsequent research will take steps to address these issues. In year 1, JIC staff began identifying relevant technology solutions to address SRTB challenges and needs; this work will provide a foundation for subsequent work focusing on assessing technology solutions in specific, real-world SRTB contexts. This work will help to build the knowledge base on technology effectiveness in the SRTB context. We also intend to build up a set of tools and resources (e.g., website, searchable database) that agencies can use to develop customized technological solutions to address both the common issues that many agencies face and individual agencies' specific needs.

Plans for Future Years

We will use the results from the advisory panel meeting to identify technologies that JIC will evaluate in its second year of funding. These evaluations will examine the costs and benefits associated with agencies' adoption of new technologies, considering acquisition costs, staff

Table 6.1
Summary of Key Findings

Method	Key Finding
Literature review	• Small, rural, tribal, and border (SRTB) agencies face many issues similar to those in urban areas, including gang activity and drug use. • SRTB agencies also face unique challenges, including long distances, low case volumes, limited staff and infrastructure, and lack of community-based experts and resources. • SRTB agencies have difficulty acquiring grant funding and training for new technologies and lack a centralized voice to affect policy and technological need assessment. • Some issues are especially challenging for certain sectors, such as the court's lack of mental health and substance-use programs and border agencies' need to address drug and human trafficking. • There is little existing research on the technology needs of SRTB agencies.
Interview	• Interviewees universally mentioned funding shortfalls, and such shortfalls colored many of the other challenges that our 12 communities of practice face. • The rarity of discretionary funding contributed to slow adoption of technology solutions and made it difficult for agencies to take advantage of economies of scale. • Geographic challenges ranged from difficulties in getting jurors and defendants to courthouses on trial dates to long transportation times for law enforcement officers responding to calls. • Information-technology (IT) challenges included antiquated case-management systems (CMSs) and jail-management systems (JMSs) and SRTB agencies' lack of technology compared with their urban counterparts. • Facility and physical infrastructure limitations were common. • Difficulties with information-sharing and interoperability across agencies were widely reported. • Language barriers between practitioners and people in the criminal justice system were common across sectors. • Tribal agencies face some unique challenges related to governance and tribal sovereignty.
Expert panel	• The majority of issues and needs discussed were not SRTB-specific but apply to larger agencies as well. • Panelists from all four agency types reported being challenged by some sort of major societal or cultural change. • Key law enforcement needs focused on improving relationships with the community, improving information-sharing, and addressing IT shortfalls. • Key court needs focused on addressing the surge in pro se litigants, improving security and resiliency at SRTB courts, and improving IT infrastructure. • Key institutional corrections needs focused on improving the provision of mental health and other services to inmates, developing the capabilities of corrections personnel, improving information-sharing through better JMSs, and addressing funding shortfalls. • Key community corrections needs focused on transitioning to an approach that emphasizes rehabilitation and positive behavioral change, improving information-sharing, addressing geographic dispersion, and addressing funding shortfalls. • Key needs across agencies included better information-sharing, assistance with procurement and management of IT systems, greater use of videoconferencing to overcome distance barriers, assistance in obtaining grants, and using nonstandard personnel to carry out certain functions at lower cost.

time, training requirements, system operation and maintenance costs, safety improvements, and performance outcomes.

Once we have selected technologies for evaluation, JIC staff will recruit agencies willing to participate in research implementations of the technology. These evaluation projects will be designed to provide important information that can be used to guide other agencies' decisions on what technology to use, what pitfalls to avoid, and what improvements they might get from the technology.

JIC will also continue to update the literature review as new sources are published and identified. JIC researchers are also contributing to a wiki that is under development at RAND; the wiki will be a searchable database that all interested parties can query to locate information on different technologies in use across various communities of practice and in different SRTB contexts. The wiki will act as a clearinghouse for information useful to practitioners and others interested in learning more about technology in criminal justice agencies.

Finally, JIC will continue to disseminate information about its work through a variety of channels. A redesigned website will be launched to highlight JIC work and will be regularly updated with information relevant to SRTB audiences. JIC researchers will continue to attend, present, and network at various conferences of relevance to SRTB criminal justice agencies. Through these conferences, JIC researchers will be able to talk about JIC work with varied audiences, present the results of its pilot research projects and other undertakings as they are completed, and keep abreast of new developments in the fields that might be of interest to practitioners. JIC researchers will also publish technology reviews through the NLECTC system as part of TechBeat.

The results of the efforts described in this volume form the foundation for continuing JIC work with the goal of fostering innovation in SRTB agencies through technology. JIC's work will guide NIJ's investments as the national focal point for work on criminal justice technology and shape the actions of other technology providers and adopters in this area. JIC will also continue to provide NIJ with guidance on where valuable TA resources should be spent to get the best value for its money. The continuing work of JIC will provide actionable guidance to SRTB agencies for prioritizing, planning, and implementing technology.

APPENDIX A

Letter of Invitation to the Study, Letter of Support for the Study, and Justice Innovation Center Announcement Letter and Mission Statement

Invitation Email (Email 1)

Subject: Request for 60-minute phone interview for Justice Innovation Center
   Dear [name]:
   My name is [name], and I am a researcher at the RAND Corporation, which is a non-profit research organization. The National Institute of Justice has funded RAND to lead the *Justice Innovation Center for Small, Rural, Tribal, and Border Criminal Justice Agencies (JIC)*. The *JIC*'s mission is to improve public safety and justice by identifying, evaluating, and disseminating technology solutions to the operational challenges of small, rural, border and tribal justice systems.
   As part of this mission, we are conducting a study to identify operational challenges that are unique to SRTB agencies or common for them. As part of this study, we are interviewing practitioners in SRTB agencies such as yours about these needs. I would like to invite you to participate in a telephone interview to discuss your perspective on these topics. For your reference, I've attached more information about the *JIC* and a letter about the study from the National Institute of Justice.
   Our questions will cover the following topical areas:

1. What are some of the biggest challenges faced by [small/rural/tribal/border] departments like your own?
2. What are some recent noteworthy technology acquisitions? What did you purchase, why did you purchase it, and how is it meeting or failing to meet your expectations?
3. What are some technologies that you want to purchase in the future and why are you interested in them? Who makes the purchasing decisions and how far out do you plan your technology acquisitions?
4. Where do you learn about technology? Is there particular information about technologies that you want, but aren't getting?

Are you free for a 60-minute phone call at any of the following dates/times? Depending on your availability, I will send you a calendar invitation with the 1-800 phone number and call details.

Dates/times (all XX time zone):

Day	Date	Times

Thank you in advance for your consideration of this request.

Confirmation Email (Email 2)

Subject: Information about upcoming phone interview with JIC

Dear [name]:

Thank you for agreeing to participate! The *Justice Innovation Center for Small and Rural Agencies (JIC)* has three goals:

1. To identify the needs of small, rural, tribal, and border agencies;
2. To identify and evaluate technologies and solutions to meet these needs;
3. To support the adoption of innovations that meet these needs in a cost-effective manner.

To help meet these goals, we are interviewing practitioners from SRTB agencies like yours to find out about your perspectives on these issues. We will be asking you questions about the challenges and unmet operational needs your agency faces, and strategies you have considered for addressing them. The questions will cover three areas:

Area 1: Your Department's Unmet Operational Needs (not necessarily related to technology)

1. What are some of the biggest challenges faced by [small/rural/tribal/border] departments like your own? We are particularly interested in the kind of challenges that might be different from those experienced by large/urban/non-border departments.
2. If the challenges you just described are not your most pressing, please describe them.

Area 2: Your Department's Technology Use

1. Please tell us about recent noteworthy technology acquisitions. What did you purchase, why did you purchase it, and how is it meeting or failing to meet your expectations?
2. What are some technologies that you want to purchase in the future and why are you interested in them? Who makes the purchasing decisions and how far out do you plan your technology acquisitions?

Area 3: How Your Department Learns about Technology
1. Where do you learn about technology? Is there particular information about technologies that you want, but aren't getting?
2. How else can we help you make informed technology decisions?

Interviews will take place over the phone and should last no more than one hour. With your permission, we would like to record the audio of the interview for note-taking purposes to ensure that we capture your response accurately. We will destroy the audio after taking notes. All information will remain completely confidential and will be aggregated when reported. We will never refer to you or your specific agency without your written approval to do so.

We look forward to speaking with you on [date].

Justice Innovation Center Announcement Letter

U.S. Department of Justice

Office of Justice Programs

National Institute of Justice

Washington, D.C. 20531

May 20, 2015

Greetings:

On behalf of the National Institute of Justice, I am pleased to announce the formation of the *Justice Innovation Center for Small, Rural, Tribal, and Border Criminal Justice Agencies*, operated jointly by the RAND Corporation and Arizona State University. The *Justice Innovation Center* is intended as a resource and a common voice for small, rural, tribal, and border (SRTB) agencies throughout the United States, which constitute the vast majority of agencies in our criminal justice system, but whose needs are often overlooked in the development and evaluation of technologies and solutions.

The mission of the *Justice Innovation Center* is to *"to improve public safety and justice by identifying, testing, evaluating, and disseminating technology solutions to the operational challenges that SRTB justice systems face."* The *Justice Innovation Center* has three main objectives: 1) to identify the challenges and unmet needs of SRTB law enforcement departments, correctional agencies, and courts; 2) to identify and evaluate technologies and solutions that address these needs in an effective and efficient manner; 3) to support the dissemination and adoption of these technologies and solutions throughout the SRTB community.

As a first step in this process, the RAND Corporation will be interviewing SRTB agencies such as yours to learn about the challenges and needs your agency is faced with, as well as your use of various technologies to address these needs. Your perspective is incredibly valuable in providing information to us about where we should direct our efforts. We encourage your participation to ensure that a wide range of voices is heard and so that we might better serve the needs of your and other SRTB agencies.

Thank you for taking the time to read this and consider our request. We look forward to hearing your feedback.

Regards,

Mike O'Shea
Senior Law Enforcement Manager
Office of Science and Technology

Justice Innovation Center Mission Statement

Justice Innovation Center for Small, Rural, Tribal, and Border Criminal Justice Agencies

Mission Statement

Innovative tactics and new technologies can drive large improvements in the effectiveness of law enforcement, corrections, and courts nationwide. Too often, however, these innovations are not designed for or never reach small, rural, tribal, and border (SRTB) justice systems.

SRTB justice systems account for three quarters of all criminal justice agencies nationwide. However, because they are so widespread and have relatively few employees, they lack a centralized voice to influence the development of technologies and other solutions.

To ensure that *all* criminal justice systems have a place at the table, the National Institute of Justice (NIJ) has developed the Justice Innovation Center for Small, Rural, Tribal, and Border Criminal Justice Agencies (JIC) with the ongoing mission:

To improve public safety and justice by identifying, evaluating, and disseminating technology solutions to the operational challenges of small, rural, tribal and border justice systems.

To achieve this mission, the JIC will bring together researchers and SRTB justice system practitioners to pursue three goals:

1. **Identify unmet operational needs that many SRTB justice systems share.** SRTB justice systems face a range of unique challenges or challenges for which the solutions available to large urban departments are impractical. The JIC will work hand-in-hand with practitioners from SRTB agencies to identify common operational requirements for which solutions are likely to contribute to large and widespread improvements in the safety, efficiency, and effectiveness of departments and agencies. It will also seek to identify types of barriers that may arise in the practical implementation and application of these technologies.

2. **Identify and rigorously evaluate new technologies or solutions to assess their effectiveness and cost-effectiveness when used by SRTB agencies.** Field tests by the IIC will consider not just whether innovations deliver needed capabilities to SRTB agencies but also their effect on operations, full life-cycle costs (including acquisition, training, and operations and maintenance), savings in time and other resources, safety issues, and how to measure their performance and demonstrate effectiveness in terms relevant at the level of the individual agency.

3. **Support the adoption of innovations that meet the operational demands of SRTB law enforcement, courts, corrections, and community corrections agencies.** The JIC will promote the exchange of innovative solutions identified by SRTB agencies spread across the United States to pool their knowledge and experience with new technologies, and will develop research-based guidance on technologies and other solutions that are proving to be effective in SRTB settings.

The RAND Corporation is a research organization that develops solutions to public policy challenges to help make communities throughout the world safer and more secure, healthier and more prosperous. RAND is nonprofit, nonpartisan, and committed to the public interest.

Santa Monica, CA | Washington, DC | Pittsburgh, PA | Boston, MA | New Orleans, LA | Cambridge, UK | Brussels, BE | Canberra, AU

APPENDIX B
Interview Coding Framework

Table B.1
Interview Coding Framework

Superordinate	Basic	Subordinate
Person-worn, training, and weapons and force	Weapons and force	
	Personnel clothing, protection or augmentation	
Vehicles	Aircraft	
	Watercraft	
	Associated technologies	
	Ground	
Technology needs and purchases	Current technology needs	Priority technology needs
	Future technology needs and purchases	Planned purchases
		Other technology of interest
	Past technology purchases, tests and inquiries	Currently employed
		No longer employed
Purchasing decision process	Deciding authority	
	Decision lead-times	
Budget information		
Sources of tech information	Help from JIC	
Nontechnology needs	Current nontechnology needs	Priority nontechnology needs
Doctrine, tactics, knowledge development, and training	Technology-mediated (or technology-enhanced) training tools	
	Tools to assist live training	
	Specialist/technologist knowledge development and training	

Table B.1—Continued

Superordinate	Basic	Subordinate
Doctrine, tactics, knowledge development, and training, continued	Societal/legal knowledge development and training	
	Management leadership knowledge development and training	
	Officer/practitioner knowledge development and training	
Facility operations technologies	Organizational logistics	
	External/perimeter physical infrastructure	
	Internal physical infrastructure	
	Internal environment control	
	Internal access control	
	Delivering services to population	
Implementation successes		
Unassigned, needs decision		
Third-party/other comments on technology		
Information and communication technologies	Information analysis	
	Information collection	
	IT, basic systems	
	Information delivery (including communications)	
	Information management (including sharing)	
Challenges	Governance and policy challenges	
	Local context issues	
	Broader societal trends	
	Operational	
	Geographical	
	Demographic	
	Financial	
	Staff	
	Technical	
Contextual information		
General comments on technology/SRTB agencies		

APPENDIX C
Members of the 2015 Justice Innovation Center Advisory Panel

Table C.1
Advisory Panel Members and Their Affiliations

Name	Position	Organization
Alex Aikman	Former court administrator	
Bob Anderson	Director	Madison County Community Corrections (Tennessee)
Danny Ball	Director	Small, Rural, Tribal, and Border Regional Center
Frank Billmayer	Under sheriff	Blaine County Sheriff's Office (Montana)
William Brueggemann	Sheriff	Cass County (Nebraska)
Joe Budnick	Chief probation officer	District 10 probation (Nebraska)
Fernando R. Castro	Deputy director	San Carlos Department of Corrections and Rehabilitation (Arizona)
Paul Chavez	Warden	Laguna Pueblo Detention Facility (New Mexico)
Thomas M. Clarke	Vice president of research and technology	NCSC
Jamie Clayton	Correctional chief	Imperial County Sheriff's Office (California)
Richard Clukey	Captain	Penobscot County Jail (Maine)
William Daly	Director	Salt River Department of Corrections (Arizona)
Jim Dennis	Executive director	Corrections Center of Northwest Ohio
Kathleen Elliott	Chief of police	Gila River Indian Community Police Department (Arizona)
Kim Ellis	Program administrator	Small, Rural, Tribal and Border Regional Center
Dalberta Faletogo	Court manager	Coeur d'Alene Tribal Court (Idaho)
William A. Ford	Supervisory physical scientist	NIJ
Lee Foster	Sheriff	Newberry County (South Carolina)
Mark Greene	Supervisory program manager	NIJ
Carla Johnson	Assistant chief	Tucson Police Department (Arizona)
Shakeba Johnson	Court administrator	Seventh Chancery District Court (Mississippi)
Kate Jones	Jail administrator	McLeod County Jail (Minnesota)

Table C.1—Continued

Name	Position	Organization
Sean Jones	Court administrator	Itasca, Koochiching, and Lake of the Woods counties (Minnesota)
Lawrence C. King	Chief judge	Colorado River Indian Tribes Tribal Court (Arizona)
Lonnie Lawson	President	Center for Rural Development
Fredric I. Lederer	Director	Center for Legal and Court Technology, College of William and Mary (Virginia)
Art Mabry	State coordinator	24/7 Sobriety Program (South Dakota)
Neil Nesheim	Area court administrator	First Judicial District (Alaska)
Rodney Olson	Court administrator	Unit 2 (North Dakota)
Michael O'Shea	Senior law enforcement program manager	NIJ
Jeff M. Pearson	Chief of police	Satellite Beach Police Department (Florida)
Joe Russo	Program manager	NLECTC
James Schneider	Director	Cass County Probation Department (Minnesota)
Steve Schuetz	Physical scientist	NIJ
Matthew Schwier	Chief of police	Chickaloon Tribal Police Department (Alaska)
Randall Soderquist	Administrator	Elko Justice and Municipal Courts (Nevada)
Matt Sparks	Sheriff	Rowan County (Kentucky)
Brett Taylor	Director of operations	Tribal Justice Exchange, Center for Court Innovation
Mark Vander Bloomen	Chief of police	Eagle River Police Department (Wisconsin)
Kim Wallace	Chief of police	City of Dover Police Department (Tennessee)
Lacie L. Wickum	Administrator	24/7 Sobriety Program (Minnesota)

APPENDIX D
Advisory Panel Read-Ahead Material

These materials, sent to advisory panel members before the meeting in December 2015, outlined the major issues and challenges identified in the interviews.

JIC Justice Innovation Center
For Small, Rural, Tribal, and Border Criminal Justice Agencies

Advisory Panel Meeting Read Ahead
December 7-8, 2015

Thank you for agreeing to participate in the advisory panel for the NLECTC Justice Innovation Center (JIC) for Small, Rural, Tribal, and Border Agencies.

Over the past year, the Justice Innovation Center (JIC) has conducted interviews with nearly 150 agencies, attended dozens of conferences, and conducted a literature review to identify major challenges facing small, rural, tribal and border (SRTB) agencies. In this meeting, we would like you to identify ways to help address these needs.

In this read ahead, we will review the format and agenda for the meeting, and summarize some of our observations to date on SRTB need for each of four criminal justice communities of practice (law enforcement, courts, institutional corrections, and community corrections).

Advisory Panel Format

After reviewing the top needs we have learned date, we will ask the advisory panel to help us identify promising *technology elements*, including physical technologies, policies and procedures, and business models that might help address the operational needs we identified. We will then ask the panel to help us identify and prioritize specific needs for technology research, development, and dissemination that would have the greatest benefits for SRTB agencies.

- On the morning of day 1, there will be brief introductions from the National Institute of Justice (NIJ) and the JIC, with the JIC's presentation summarizing the top operational issues emerging from the agency interviews. Panelists will then introduce themselves, briefly giving an example of both a technology acquisition they are familiar with that did not work well and an acquisition that did work well (or at least seems promising).
- From mid-morning on day 1 to the end of day 1, panelists will split into four breakout groups (law enforcement, courts, community corrections, and institutional corrections) to discuss top issues facing their field, promising technology elements.
- The morning of day 2 will start with summaries of the top issues identified in each breakout group. The full group will discuss the breakout groups' results, and whether there are priority needs that cut across the criminal justice system.
- On mid-morning on day 2, the panelists will fill out an electronic questionnaire to assess the likely benefits to SRTB agencies of identifying practical solutions to each of the priority needs.
- Finally, during the late morning of day 2, the panelists will discuss two special topics for the JIC – setting up an ongoing panel survey of SRTB practitioners and identifying what information products would be most useful for the JIC to develop for SRTB practitioners.

Agenda

Day 1 Monday, December 7

9:00 Introduction and Overview (National Institute of Justice / RAND)

10:00 Introduction of Panel Members

11:00 Break Out Groups (Law Enforcement, Courts, or Corrections)

12:00 Break for Lunch

1:00 Break Out Groups (continued)

5:00 Adjourned

Day 2 Tuesday, December 8

9:00 Summary of Break Out Group Issues and Technology Solutions

9:45 Group Discussion of Cross-Sector Issues and Technologies

10:30 Electronic Panelist Assessment of Solutions

11:15 Discussion of special topics for the JIC:

- Ongoing panel survey of SRTB practitioners
- What would make the JIC website most useful

12:00 Meeting Adjourned

Top Operational Issues from Prior JIC Research

Below, we have summarized the top issues and themes from our interviews with dozens of criminal justice practitioners, as well as our reviews of prior literature related to small, rural, tribal, and border agencies. Results are presented in the following order: law enforcement, courts, institutional corrections, and community corrections. The law enforcement discussion includes breakouts specifically for tribal and border agency issues.

Broadly speaking, we identified four top-level themes in the issues. These include:

- Shortfalls related to *information technology*, including interoperability, infrastructure, and difficulties upgrading to new systems.
- Shortfalls related to *agency operations*, including policies, governance, "business" processes, training, recruiting, and retention.
- Shortfalls specifically due to the geographic dispersion of many SRTB agencies, including the challenges to having to cover large areas, travel long distances regularly, and having comparatively few services and resources locally.
- Shortfalls due to lack of funding and other resources that can be especially acute for SRTB agencies.

Law Enforcement Needs
Information Technology Challenges

- **Interoperability.**
    - Radio communications interoperability across agencies and with other first responder agencies is lacking.
    - Data interoperability across agencies (e.g., with records management system and federal repository data) is also challenging.
    - Accessing key information needed in the field is difficult, in large part due to a lack of infrastructure (see below).
- **Special interoperability issues for border agencies.**
    - Communications with partnering agencies across borders can be limited.
    - Communications towers and rules in neighboring countries can cause interference.
    - Border agencies may not have personnel with sufficient clearances to work with the FBI and other federal agencies whose work often brings them border regions.
- **Infrastructure.**
    - Communications radio technology can be antiquated and /or lacking necessary equipment.
    - Other common field technology deficits include a lack of laptop computers and cameras in patrol cars, license plate readers, TASERs, and radios.
- **Impacts of large areas and distances.**
    - Distances can make radio frequencies weak.
    - Distance, terrain features, and a lack of commercial cell coverage in large areas, can cause large gaps in coverage.
- **Difficulties in adopting new technologies.**
    - Agencies suffer from a general shortfall in IT personnel capable of acquiring, installing, and maintaining core systems.
    - Technologies developed for larger departments are not always applicable to smaller agencies.

Agency Operations Challenges

- **Lack of specialization.** In small agencies, staff must become generalists for both the full range of law enforcement operations as well as technologies.
- **Difficulties in recruiting and retention.**
    - Finding qualified candidates and retaining them is challenging, in part due to salary and equipment competition from larger departments.
    - Chronic understaffing and overwork can also result in burnout.
    - Finding multilingual officers can be especially challenging.
- **Difficulties in relationships with the community.**

- o Police-community relations problems (i.e., lack of trust) have created communication issues with community residents.
  - o Language barriers have limited the amount of communication between the community and law enforcement, in both directions.
- **Issues specifically for border agencies**.
  - o Border agencies frequently struggle to establish trustworthy relationships with international law enforcement agencies.
  - o Police involvement in immigration issues can further degrade community-police relations.
  - o Language and cultural barriers between the police and immigrant communities can be particularly acute.
- **Issues specifically for tribal agencies**. Information sharing and joint operations between tribal and non-tribal agencies are especially problematic due to tribal sovereignty and cultural issues.
- Tribal locations do not necessarily follow traditional US postal service rules for addresses, making finding locations of calls and incidents and resulting data quality challenging.

Geographic Challenges

- **Lack of key local resources**. Juvenile detention facilities and jails may not be available locally, requiring officers to spend large portions of their shifts driving arrestees to facilities that are hours away. This travel time takes officers off-duty for much of their shifts.
- **Impacts of large distances**. Agencies may be challenged by having to cover large swaths of territory, with patrols and responses separated by hours of driving.

Funding and Resourcing Challenges

- **General**. There is a lack of funding in general, to include being able to maintain sufficiently-sized law enforcement agencies needed to provide sufficient patrol levels, perform crime analysis, and train to maintain skills. The budget cutbacks post-2008 have made staffing and other shortfalls acute, especially in areas with high rates of substance abuse, domestic violence, drugs trafficking and human trafficking. The Attorney General's recent limitations on using confiscated assets for funding have reduced budgets. Unfunded mandates, such as some states' requirements that police use body worn cameras, have put further pressure on budgets.
- **Technology funding**.
  - o Funding to acquire or maintain key equipment, including radio communications systems and other key pieces of technology identified under "Infrastructure" above, is extremely limited.
  - o Body worn cameras were seen as being especially expensive, in terms of purchasing the cameras and managing the resulting video footage.

- o Limited funding also precludes staff from being able to attend technology conferences and training classes.
- o Because smaller agencies cannot buy in bulk, procurement costs can be more expensive per piece of equipment than in larger agencies.
- **Federal funding.**
  - o Respondents felt they were unlikely to compete successfully for federal technology funding against large agencies.
  - o Agencies that win grant funding for new systems are then challenged by the long term costs of operating and maintaining that system.
- **Special issues for border agencies.**
  - o Border agencies can be especially challenged by high volumes of drug and human trafficking.
  - o There are concerns that cuts to border security grants (Stonegarden, etc.) from DHS will reduce operational capacity.

Courts Needs
Information Technology Challenges

- **Interoperability shortfalls**.
    - A lack of sharing infrastructure such as authentication/ permission control and flexible license agreements prevent judges and qualified staff from electronic access to important case records.
    - Case management systems for civil, family, and criminal courts are antiquated or dysfunctional.
    - Components that should work together are incompatible (e.g. financial system is divorced from case management systems).
    - Some electronic systems lack interoperability for sharing information with external partners (e.g. police, probation).
- **Infrastructure shortfalls**.
    - Communications infrastructure (broadband, cellphone coverage) is unreliable or unavailable and may hinder adoption of technological solutions (e.g. GPS monitoring).
    - Even basic utilities such as electricity can be unreliable.
    - There also is a lack of web infrastructure necessary to provide services to the public in some places.
- **Difficulties of adopting technology**
    - Systems can be difficult to adapt to statutory and policy changes.
    - Many courts retain manual, paper-based workflows. Manual filing of forms can cause confusion and delay among the increasing number of self-represented litigants. Manual records management, payment, reporting, and payout increases caseloads due to the heavy time investment required of staff. Digitization efforts are often behind schedule, because they are manually intensive or costly.
- **Language barriers**.
    - Courts lack adequate interpreter and translation services for the increasing number of non-English speaking clients.
    - Telephonic translation services may be inappropriate for longer proceedings.

Agency Operations Challenges

- **Process innovation**.
    - Dispute resolution procedures are inefficient compared to innovative practices used by online competitors and alternate dispute resolution providers. This is bad for ensuring that parties needing the courts use them and delays access to justice.
    - Courts may be slow to adopt innovations due to a culture of distrust of technology among many judges and other influential figures.

- o Some jurisdictions are hampered by political opposition to alternatives to incarceration (e.g., treatment).
- o Technology standards for evidence are lacking.
- **Access to justice**.
  - o A lack of accessibility (such as mobile web optimization, use of plain language, and general user-friendliness) hinder clients' attempts to participate in justice procedures.
  - o Courts are unprepared to handle the increasing *pro se* population.

Geographic Challenges

- **Distance challenges**.
  - o Lengthy travel time limits timely access to resources (e.g., IT support, training, treatment, reentry services) and delays implementation of change in outlying offices.
  - o Many activities require physical presence at courthouse (e.g. original signature on court documents) due to legal rules, which requires lengthy travel times for all involved parties. This can limit the number of private bar attorneys willing to take cases in remote courts.
  - o Public transportation is limited, which can be especially challenging for the elderly, those with license revocations (e.g. DUI probationers), and children who are involved in proceedings.
- **Shortfalls in local capabilities in rural areas**. At a local level, rural areas experience severe limitations in: training opportunities and IT staff; treatment programs, service providers for mental health and substance use, and other local reentry/reintegration services (e.g. halfway houses, counseling); and legal service providers (e.g., lawyers, bonding agencies).

Funding and Resourcing Challenges

- **Revenues**. Courts suffer from budget instability due to broad socioeconomic trends (e.g. recession, unreliable tax revenue, declining civil caseloads). They can be over-reliant on specific industries for tax revenue (e.g. coal, oil). The declining working-age population in some areas also reduces tax revenue.
- **Equipment**.
  - o Small courts lack the resources to invest in sophisticated technologies.
  - o Many courts lack basic security measures (e.g. security staff, barriers, screening technologies, cameras).
- **Staffing**.
  - o Small courts are unable to benefit from specialization of courts and judges (e.g. problem-solving courts).

- When a few judges are covering multiple courts, this can lead to different judges handling the same case, reducing continuity.
- Small courthouses can mean the court shuts down if a single person is sick, at least until someone else can be found to fill in for them (but that person may be an hour or more away at a neighboring county court).

Institutional Corrections Needs
Information Technology Challenges

- **Interoperability.** Data, including records from court agencies, law enforcement and neighboring jail and prison facilities—necessary for better identifying inmate needs and risks—are difficult for jail staff to obtain.
- **Infrastructure.**
  - Many agencies lack sufficient bandwidth to run systems including RMS, video visitation, telemedicine and remote arraignment. These problems lead to increased transportation and contracting costs, as well as added stress to staff and inmates as services must be conducted offsite.
  - Slow internet speeds hamper opportunities to offer inexpensive internet-based services such as job training and legal research to inmates.
  - Wiring in older facilities is not sufficient to support digital cameras and improved jail management systems.
- **Jail Management Systems.**
  - JMS platforms are often not user friendly, do not function correctly, or cannot be used to full capacity with many older jails' current infrastructure. Many JMSs are based on law enforcement needs, and were not designed with corrections tasks in mind.
  - Redundancy in data entry and needless complexity are common, and customization requires programming expertise that many agencies lack.
- **Qualified Vendors.**
  - Reliable, ongoing support from vendors and the cost of maintenance are challenges. Vendors may understate costs or fail to provide long-term support.
  - Some areas suffer from limited vendor options because of their remote locations, or are required to deal with local vendors who are less competent than larger, more experienced vendors.

Agency Operations Challenges

- **Mental Health and Substance Abuse Treatment.**
  - A large number of inmates suffer from chemical dependency issues, physical health problems, and especially mental health problems, and correctional agencies lack sufficient resources to address the problems.
  - In agencies that cannot afford to keep social workers or nurses on staff inmates requiring care must be transported to external facilities, taking up valuable staff time and resources.
  - Reliance on external providers limits the ability of correctional facilities to respond to inmates in crisis.

- o   Limited aftercare or reentry support services for released inmates increase the likelihood that an inmate will return to confinement.
- **Staffing.**
  - o   Many agencies face long-term problems due to a lack of IT management expertise.
  - o   Turnover makes it difficult to create institutional knowledge about IT systems and decisions.
  - o   In some states, retaining staff is difficult because being a correctional officer is no longer considered a stepping stone to becoming a law enforcement officer. In other areas, staff are hard to retain because after gaining some experience, staff leave to become law enforcement officers.

Geographic Barriers

- **Transportation.**
  - o   Geographic isolation limits agencies' access to potential staff, treatment services, community interaction, and visitation for inmates.
  - o   In areas where the court system does not offer video arraignment, transportation of inmates to court, often for very short hearings, can take significant amounts of staff time and coordination.

Funding and Resourcing Challenges

- **Budget.**
  - o   Funding often lags need for institutional corrections, leading to low wages and slow adoption of new technology, and in turn difficulties attracting and retaining qualified staff.
  - o   Agencies frequently lack discretionary funds for technology acquisition. In some states, technology planning is handled at the state level and new equipment is implemented by mandate, reducing administrator decision-making power.
  - o   Diverse funding sources (local, regional, state, federal) require significant time and resources be spent on procurement and acquisition processes.
- **Facilities.** Older facilities present challenges for staff that are exacerbated by Prison Rape Elimination Act (PREA) mandates, and changes in solitary confinement and segregation policies. In older facilities, challenges with getting into compliance with PREA—and improving surveillance in general—is complicated by poor sightlines, narrow corridors, isolated areas, low ceilings, and outdated wiring. Those in very old facilities may forgo technology improvements until a new facility is built.
- **Addressing Health Needs.** Telemedicine and remote treatment services are the most promising approach to addressing significant inmate health needs, but they come with high start-up costs—financially and otherwise. Many facilities lack the physical space for

and funding to purchase required equipment. Coordinating with providers and insurance companies can be a huge barrier to implementing telemedicine.

Community Corrections Needs
Information Technology Challenges

- **Interoperability challenges.** Limited data sharing with other agencies in the same state is a common and significant challenge. Data from agencies using mandated state systems are typically kept in county- or district-level silos. Other states allow each county to choose its own case management system, limiting the interoperability of systems between jurisdictions. Limitations on data sharing among different agencies regarding active clients reduce the ability of officers to determine client risk level.
- **Infrastructure challenges.**
    - Some rural and border agencies struggle to support basic technologies, considered standard in larger or urban agencies.
    - Limited cell phone coverage and WiFi access are common in extremely rural areas. Cell phone coverage limits the ability of some agencies to use remote monitoring devices.
    - Officers sometimes have limited access to mobile devices because of cost and challenges with connectivity.
    - Agencies lack sufficient systems to track and collect supervision fees from clients.
- **IT Support.** Small and rural agencies usually do not have the resources to support their own IT departments and rely on the county-level information technology departments to provide IT support to their offices. The ability of county IT offices can be limited in these areas and community corrections departments must compete with other county agencies and priorities for IT support. County IT departments may also have limited understanding of correction agencies' needs.
- **Remote data access.** Necessary access to client data for field officers can be challenging. Remote data access is complicated by reliance on a CMS that is not web-based or accessible through any web-browser, or a lack of apps created by the software vendor to allow remote access to the data system through a smart phone or tablet. Security concerns can further limit the ease with which field officers access data remotely.

Agency Operations Challenges

- **Electronic files.** Agencies have limited ability to go paperless or limit their use of paper forms because of reporting and filing mandates from other agencies, including state-level oversight agencies and court systems. Agencies lack access to or the ability to use electronic forms, electronic signatures, and e-filing of court documents, like presentence investigations.
- **Substance testing.**
    - Rapid drug tests using urine samples, used by most community corrections agencies, can be time consuming for staff and clients.

- o Many agencies must rely on local law enforcement to conduct tests using their portable breath testers, which are more accurate but prohibitively expensive for community corrections agencies.
- **Office management.**
    - o Many agencies lack electronic systems that allow clients to check-in for meetings at the office in order to reduce the need for an administrative assistant to spend time interacting with clients and notifying officers about their schedule.
    - o More limited record keeping systems regarding client meetings, payments, and reporting requirements reduce accountability on the part of the officer and the client.

Geographic Challenges

- **Transportation.**
    - o In agencies where home visits are required—especially among those with higher-risk caseloads—getting to clients and getting clients to services poses significant challenges.
    - o Conducting home visits or employment checks and traveling to local courts across large districts reduces the caseload that each staff member can carry, increasing staff needs.
    - o Getting clients to services that are far away is complicated by limited public transportation and the high costs of transportation like cabs. Difficulty with transportation reduces clients' likelihood for successfully completing supervision.

Funding and Resourcing Challenges

- **Budget.**
    - o Funding for technology upgrades is extremely limited.
    - o Financial constraints are especially significant for agencies operating at the county level, where supportive or adversarial county boards can mean the difference between regular technology upgrades and repeating requests for technology over several years before receiving approval.
    - o Limited resources restrict the number of staff agencies can hire or the amount of overtime officers can incur.
- **Service Provision.**
    - o The availability of high quality treatment programs and support services in many rural areas limited.
    - o Agencies frequently have difficulties getting clients into services, whether due to the high cost of the program that has to be paid by the probationer, the lack of programs available in rural areas, or the lack of openings in existing programs.
    - o Increased heroin use is creating an increased demand for drug treatment services.

- Specialized service provision and treatment, such as anger management treatment or shoplifting programs, are difficult to find and difficult to place clients into. Appropriate online programs or programs provided over remote video links are limited.

APPENDIX E

Tier 1 Needs from the Justice Innovation Center Advisory Panel, by Agency Type

Table E.1
Agency Types, Issues, and Recommendations: Geography

Agency Type	Issue	Recommendation
Law enforcement	Agencies suffer from a general shortfall in IT personnel capable of acquiring, installing, and maintaining core systems.	Assess the feasibility of giving approved vendors remote access.
Court	At a local level, rural areas experience severe limitations in training opportunities and IT staff; treatment programs, service providers for mental health and substance use, and other local reentry and reintegration services (e.g., halfway houses, counseling); and legal service providers (e.g., lawyers, bonding agencies).	Provide access to reliable videoconferencing that is simple and works every time. Evaluate the utility and feasibility of remote mental health evaluations and remote probation reports.
	Lengthy travel time limits timely access to resources (e.g., IT support, training, treatment, reentry services) and delays implementation of change in outlying offices.	Evaluate the cost-effectiveness of telemedicine mental health, substance use, and other behavioral health services in rural areas that have no local mental health services.
Institutional corrections	Geographic isolation limits agencies' access to potential staff, treatment services, community interaction, and visitation for inmates.	Research is needed on implementation and best practices for use of telemedicine (separate from telepsychiatry); JIC can also take lessons from the use of telemedicine outside of the corrections arena.
Community corrections	Conducting home visits or employment checks and traveling to local courts across large districts reduces the caseload that each staff member can carry, increasing staff needs.	Establish an in-person relationship first, then leverage remote technologies.
	In agencies that require home visits—especially among those with higher-risk caseloads—getting to clients and getting clients to services pose significant challenges.	Use videoconferencing for telehealth, medical, and mental health for home visits (might require legal or policy changes), with verified confidentiality and security.

Table E.2
Agency Types, Issues, and Recommendations: Information Technology

Agency Type	Issue	Recommendation
Law enforcement	Data interoperability across agencies (e.g., with RMSs and federal repository data) is challenging.	Assess the feasibility of developing state and national standards to share data.
	The NIBRS-in-five-years conversation (a mandate that a department must submit all data to NIBRS within five years) needs to take place.	Develop models to become NIBRS-compliant.
	Distances can make radio frequencies weak.	Upgrade analog systems to digital systems.
	Communication radio technology can be antiquated or lack necessary equipment.	Assess the viability of a web-based system for trading and buying technology between law enforcement agencies (e.g., eBay for law enforcement equipment).
	Data interoperability across agencies (e.g., with RMSs and federal repository data) is challenging.	Assess the scalability of developing patches to allow systems to communicate.
	Radio communication interoperability across agencies and with other first-responder agencies is lacking.	Work with the FCC to ensure consistency and provision of needed radio spectrum to SRTB agencies.
Court	There is a lack of web infrastructure necessary to provide services to the public in some places.	Governments with inadequate broadband services should develop improved broadband capabilities even if they must be shared across multiple local and state agencies. Evaluate the cost-effectiveness of using satellite uplinks as an Internet and telephone solution in areas without cell phone service.
	Courts lack adequate interpreter and translation services for the increasing number of non–English-speaking clients.	Interpreters who are certified in other fields (e.g., health care or diplomacy) need simplified court certification processes that will encourage more to obtain court certification at lower cost.
	Privacy issues arise when records are digitized.	When developing new CMSs, states and courts need to plan in advance what information will be shared with the public (or others). Model policies on information-sharing, such as those that NCSC developed, should be evaluated and disseminated. (Note: In the discussion, there was not a clear assignment of who should do this.)
	Communication infrastructure (broadband, cell phone coverage) is unreliable or unavailable and might hinder adoption of technological solutions (e.g., GPS monitoring).	Assess the utility of low-cost general-purpose technology systems (such as Skype) for SRTB courts to use rather than purchasing customized systems.
		Assess the effectiveness and legality (or required changes to law) of using unconventional strategies for communicating with remote populations, such as use of Facebook or texting.
		Assess new systems and technology that rural courts could use to exploit First Responder Network Authority (FirstNet) capabilities.
	Courts need to see new technologies working to have confidence in them.	Establish or identify model or mentor SRTB courts demonstrating processes and technologies. The status could be competitively awarded to support visitors. This need applies to a range of court types.

Table E.2—Continued

Agency Type	Issue	Recommendation
Institutional corrections	JMS platforms are often not user-friendly, do not function correctly, or cannot be used to full capacity with many older jails' current infrastructure. Many JMSs are based on law enforcement needs and were not designed with corrections tasks in mind.	National organization should develop template for vendors to become JMS certified as experts in corrections management. A national organization (e.g., NIJ, ACA) needs to develop checklist of mandatory elements that should be in a JMS, including the ability to generate required data for state and federal reporting and visual mapping of in-facility hot spots.
	Reliable, ongoing support from vendors and the cost of maintenance are challenges. Vendors might understate costs or fail to provide long-term support.	ACA or BJA should make IT experts available for consulting on IT specific to the corrections field.
Community corrections	Necessary access to client data for field officers can be challenging. Remote data access is complicated by reliance on a CMS that is not web-based or accessible through any web browser or a lack of apps created by the software vendor to allow remote access to the data system through a smartphone or tablet. Security concerns can further limit the ease with which field officers access data remotely.	Provide tablet or phablet (large cell phone) with integrated CMS and other work task software and remote access to important databases.
	There are many interoperability challenges. Limited data-sharing with other agencies in the same state is a common and significant challenge. Data from agencies using mandated state systems are typically kept in county- or district-level silos. Other states allow each county to choose its own CMS, limiting the interoperability of systems between jurisdictions. Limitations on data-sharing regarding active clients among different agencies reduce officers' ability to determine client risk level.	Build a statewide (and potentially broader) software platform that aggregates disparate systems to link agencies and allows selective access to data based on agency needs and requirements. This will integrate an alert system for immediate issues (e.g., rearrest) from other agencies. This will involve appropriate MOUs.
	Some rural and border agencies struggle to support basic technologies that are considered standard in larger or urban agencies.	Implement access-control systems to allow Internet access for work tasks and maintain network security.
	Officers sometimes have limited access to mobile devices because of cost and challenges with connectivity.	Provide every officer with a tablet or laptop with a data connection.

Table E.3
Agency Types, Issues, and Recommendations: Operations

Agency Type	Issue	Recommendation
Law enforcement	Police–community relations problems (i.e., lack of trust) have created communication issues with community residents.	Work with the local press to announce positive developments.
		Develop a social media app through which the public can provide feedback.
		Identify strategies for increasing agency presence in the community.
		Increase reliance on community surveys to obtain feedback on agency performance.
	Technologies developed for larger departments do not always apply to smaller agencies.	Develop protocols to help small agencies more precisely determine their IT needs and to match those needs with IT vendor products.
	In small agencies, staff must become generalists for both the full range of law enforcement operations and technologies.	Identify and develop a low-cost solution to growing internal IT capacity.
Court	Courts are unprepared to handle the increasing pro se population.	Develop TurboTax–like form completion for pro se litigants (or the public).
		Develop and distribute how-to videos in the courthouse and on YouTube explaining procedures.
	Courts often provide little or only ad hoc training to staff.	Provide online training resources for court staff, such as what Utah has developed (short videos and tests).
		Provide scholarships for court staff to attend trainings.
	Explaining court procedure to the public is challenging.	Develop or identify model resource guides, brochures, and website content for educating the public on court procedures.
	There is a need for disaster recovery plans, including robust backup systems for court IT systems.	Courts need continuity-of-operations plans and planning tools.

Table E.3—Continued

Agency Type	Issue	Recommendation
Institutional corrections	A large number of inmates suffer from chemical dependency issues, physical health problems, and especially mental health problems, and corrections agencies lack sufficient resources to address the problems.	At a minimum, jails should provide staff with training in deescalation, and deescalation training should be done at academy.
		Agencies should develop good relationships with mental and behavioral health agencies.
		Mandates at the state level should require other agencies to work with jails on interacting with people who have mental illnesses.
		Jails should use critical-incident and crisis-intervention teams to manage inmates in crisis.
		Agencies need information on what the best practices are for the implementation and use of telepsychiatry.
		Create partnerships between law enforcement, jails, mental health, and health and human service agencies to address placement for people with mental health problems.
		Agencies need a more proactive approach to mental health, including a good assessment tool that they can use before admission to identify mental health needs and help make decisions about care and placement.
		Agencies should analyze inmate data to understand the effect that mentally ill inmates have on facility operations (e.g., the revolving door).
	Turnover makes creating institutional knowledge about IT systems and decisions difficult.	Agency heads should attend to the development of their line staff, make them feel heard, and involve them in committees to increase their investment in the agencies.
		Academy training needs to focus specifically on corrections and not law enforcement.
		Agencies should provide continued training on deescalation and self-defense for staff so they feel safe.
	Limited aftercare or reentry support services for released inmates increase the likelihood that an inmate will return to confinement.	Agencies should create a handbook of resources in community (e.g., housing, employment) to provide to inmates upon release.
	Accreditation processes are insufficient.	Agencies should use ACA standards to measure their facilities, identify weaknesses, and see where improvement is needed. Even if they are not going after accreditation, these standards can provide information on best practices.

Table E.3—Continued

Agency Type	Issue	Recommendation
Community corrections	Many agencies must rely on local law enforcement to conduct tests using their portable breath testers, which are relatively accurate but prohibitively expensive for community corrections agencies.	Maintain good, if informal, relations with law enforcement for reciprocal support, particularly in remote areas, where coverage is thinner.
	Limited record-keeping systems regarding client meetings, payments, and reporting requirements reduce officer and client accountability.	Redefine the essential role of the probation officer as a client-focused hybrid of law enforcement and social worker, and eliminate tasks that fall outside that scope.
	Limited record-keeping systems regarding client meetings, payments, and reporting requirements reduce accountability on the part of the officer and the client.	Identify and prioritize high domains in risk assessments, then link needs to research-based treatment programs.
	The changing nature of community corrections requires an approach that is less punitive and more social work oriented.	Foster interactions for role-modeling behavior in the community.
	Agencies have a limited ability to go paperless or limit their use of paper forms because of reporting and filing mandates from other agencies, including state-level oversight agencies and court systems. Agencies lack access to or the ability to use electronic forms, electronic signatures, and e-filing of court documents, such as PSI reports.	Eliminate antiquated bureaucratic processes that require paper forms or local hard-copy storage or that prohibit scanning.

Table E.4
Agency Types, Issues, and Recommendations: Funding and Resourcing

Agency Type	Issue	Recommendation
Law enforcement	There is a lack of funding in general, including funds to maintain sufficiently sized law enforcement agencies needed to provide sufficient patrol levels, perform crime analysis, and train to maintain skills. The budget cutbacks post-2008 have made staffing and other shortfalls acute, especially in areas with high rates of substance use, domestic violence, drug trafficking, and human trafficking. The attorney general's recent limitations on using confiscated assets for funding have reduced budgets. Unfunded mandates, such as some states' requirements that police use BWCs, have put further pressure on budgets.	Sharpen definitions of small agencies, and develop organizational mechanisms to build grant-development capacity. Develop TA centers for assisting agencies with funding applications.
	Funding to acquire or maintain key equipment, including radio communication systems and other key pieces of technology identified under "Infrastructure" above, is extremely limited.	Research and identify options for financing capital equipment.
	BWCs are seen as being especially expensive, in terms of purchasing the cameras and managing the resulting video footage.	Provide small agencies with better information on the acquisition, use, and policy development and long-term cost of adopting BWCs.
	Agencies might be challenged by having to cover large swaths of territory, with patrols and responses separated by hours of driving.	Develop a model MOU for interagency mutual aid.
	Limited funding also precludes staff from being able to attend technology conferences and training classes.	Develop distance-learning strategies for small and rural agencies.
	Because smaller agencies cannot buy in bulk, procurement costs can be higher per piece of equipment than for larger agencies.	Explore leveraging state procurement programs to save money.
Courts	Small courts lack the resources to invest in sophisticated technologies.	Provide TA services funded explicitly for SRTB courts.
	Many courts lack basic security measures (e.g., security staff, barriers, screening technologies, cameras).	Provide emergency response training to court staff. Establish widespread grading and evaluation of security at courthouses.
	Funding in general is scarce.	Government grant programs should encourage broad and innovative uses of purchased technology, not limiting purchases to single uses.
Institutional corrections	Use of diverse funding sources (local, regional, state, federal) requires that significant time and resources are spent on procurement and acquisition processes.	Federal grants should allow funding for a part-time employee to administer grants.

Table E.4—Continued

Agency Type	Issue	Recommendation
Community corrections	The availability of high-quality treatment programs and support services is limited in many rural areas.	Catalog available treatment facilities and programs in neighboring areas, including capacity, cost, and effectiveness.
	Funding for technology upgrades is extremely limited.	Provide support in identifying new grant opportunities and writing grant submissions.
		Assess feasibility, and plan for life-cycle costs of potential grant-funded programs.
	Specialized service provision and treatment, such as anger-management treatment or antishoplifting programs, are difficult to find and difficult to place clients into. Appropriate online programs or programs provided over remote video links are limited.	Leverage program performance information that monitoring technologies generate into positive reinforcements.
	Limited resources restrict the number of staff agencies can hire or the amount of overtime officers can incur.	Use support staff for nontherapeutic needs.
		Use dictation software for fillable forms in the field.
		Routinely revisit data collection to minimize redundancy and verify that the data collected are useful.

Bibliography

ABA—*See* American Bar Association.

AJA—*See* American Jail Association.

American Bar Association, Standing Committee on Pro Bono and Public Service, and the Center for Pro Bono, *Rural Pro Bono Delivery: A Guide to Pro Bono Legal Services in Rural Areas*, Chicago, Ill., 2003. As of August 22, 2016:
http://www.americanbar.org/content/dam/aba/images/probono_public_service/ts/aba_rural_book.pdf

American Jail Association, "Statistics of Note," undated. As of July 5, 2016:
https://members.aja.org/About/StatisticsOfNote.aspx

———, *Jail Directory Reporter*, March 2015.

Applegate, Brandon K., and Alicia H. Sitren, "The Jail and the Community: Comparing Jails in Rural and Urban Contexts," *Prison Journal*, Vol. 88, No. 2, June 2008, pp. 252–269.

Arnold, Aaron F., Sarah Cumbie Reckess, and Robert V. Wolf, "State and Tribal Courts: Strategies for Bridging the Divide," *Gonzaga Law Review*, Vol. 47, No. 3, 2012, pp. 801–838.

Atherton, Gene, and Joe Russo, "NIJ's Technical Assistance for Corrections," *Corrections Today*, August 2009, pp. 48–50. As of August 22, 2016:
https://www.ncjrs.gov/pdffiles1/nij/249039.pdf

Beals, Janette, Douglas K. Novins, Nancy R. Whitesell, Paul Spicer, Christina M. Mitchell, and Spero M. Manson, "Prevalence of Mental Disorders and Utilization of Mental Health Services in Two American Indian Reservation Populations: Mental Health Disparities in a National Context," *American Journal of Psychiatry*, Vol. 162, No. 9, September 2005, pp. 1723–1732.

Bean, Philip, "Technology and Criminal Justice," *International Review of Law, Computers, and Technology*, Vol. 13, No. 3, 1999, pp. 365–371.

BIA—*See* Bureau of Indian Affairs.

BJS *See* Bureau of Justice Statistics.

Black, M. C., K. C. Basile, M. J. Breiding, S. G. Smith, M. L. Walters, M. T. Merrick, J. Chen, and M. R. Stevens, *The National Intimate Partner and Sexual Violence Survey (NISVS): 2010 Summary Report*, Atlanta, Ga.: National Center for Injury Prevention and Control, Centers for Disease Control and Prevention, 2011. As of August 26, 2016:
https://www.cdc.gov/violenceprevention/pdf/nisvs_report2010-a.pdf

Breiding, Matthew J., Sharon G. Smith, Kathleen C. Basile, Mikel L. Walters, Jieru Chen, and Melissa T. Merrick, "Prevalence and Characteristics of Sexual Violence, Stalking, and Intimate Partner Violence Victimization: National Intimate Partner and Sexual Violence Survey, United States, 2011," *Surveillance Summaries*, Vol. 63, SS. 8, September 5, 2014, pp. 1–18. As of August 22, 2016:
https://www.cdc.gov/mmwr/preview/mmwrhtml/ss6308a1.htm

Brown, Mary Maureen, "The Benefits and Costs of Information Technology Innovations: An Empirical Assessment of a Local Government Agency," *Public Performance and Management Review*, Vol. 24, No. 4, June 2001, pp. 351–366.

Bureau of Indian Affairs, Division of Law Enforcement, "Operations," last updated March 8, 2013. As of August 22, 2016:
http://bia.gov/WhoWeAre/BIA/OJS/DOLE/

Bureau of Justice Statistics, Office of Justice Programs, U.S. Department of Justice, "Data Collection: National Crime Victimization Survey (NCVS)," undated. As of September 12, 2016:
http://www.bjs.gov/index.cfm?ty=dcdetail&iid=245

———, *Law Enforcement Management and Administrative Statistics (LEMAS), 2013*, Ann Arbor, Mich.: Inter-university Consortium for Political and Social Research, ICPSR36164-v2, September 22, 2015a. As of August 26, 2016:
http://doi.org/10.3886/ICPSR36164.v2

———, *Annual Survey of Jails, 2014*, Ann Arbor, Mich.: Inter-university Consortium for Political and Social Research, October 12, 2015b. As of August 23, 2016:
http://doi.org/10.3886/ICPSR36274.v1

Burruss, George W., Joseph A. Schafer, Matthew J. Giblin, and Melissa R. Haynes, "Homeland Security in Small Law Enforcement Agencies: Preparedness and Proximity to Big-City Peers," *NIJ Journal*, No. 274, December 2014, pp. 1–4. As of September 12, 2016:
https://www.ncjrs.gov/pdffiles1/nij/247882.pdf

Byrne, James, and Gary Marx, "Technological Innovations in Crime Prevention and Policing: A Review of the Research on Implementation and Impact," *Cahiers Politiestudies Jaargang*, Vol. 3, No. 20, 2011, pp. 17–40. As of August 22, 2016:
https://www.ncjrs.gov/pdffiles1/nij/238011.pdf

Center for Court Innovation, "Tribal Justice," undated. As of August 26, 2016:
http://www.courtinnovation.org/topic/tribal-justice

Champagne, Duane, and Carole Goldberg, *Promising Strategies: Public Law 280*, Washington, D.C.: U.S. Department of Justice, Bureau of Justice Assistance, March 2013. As of August 22, 2016:
https://www.ncjrs.gov/app/publications/abstract.aspx?ID=265859

Cobb, Kimberly A., with Mary Ann Mowatt, Adam Matz, and Tracy Mullins, *A Desktop Guide for Tribal Probation Personnel: The Screening and Assessment Process*, Washington, D.C.: U.S. Department of Justice, Bureau of Justice Assistance, May 2011. As of August 22, 2016:
https://www.bja.gov/Publications/APPA_TribalProbation.pdf

Cobb, Kimberly A., with Mary Ann Mowatt and Tracy Mullins, *Risk-Needs Responsivity: Turning Principles into Practice for Tribal Probation Personnel*, American Probation and Parole Association, August 2013. As of August 22, 2016:
https://www.bja.gov/Publications/APPA-RNR-Tribal-Probation.pdf

Cobb, Kimberly A., and Tracy G. Mullins, "Tribal Probation: An Overview for Tribal Court Judges," *Journal of Court Innovation*, Vol. 2, No. 2, Fall 2009, pp. 329–344.

———, *Tribal Probation: An Overview for Tribal Court Judges*, Washington, D.C.: U.S. Department of Justice, Bureau of Justice Assistance, May 2010. As of August 22, 2016:
https://www.appa-net.org/eweb/docs/appa/pubs/TPOTCJ.pdf

Coburn, Andrew F., A. Clinton MacKinney, Timothy D. McBride, Keith J. Mueller, Rebecca T. Slifkin, and Mary K. Wakefield, *Choosing Rural Definitions: Implications for Health Policy*, Rural Policy Research Institute Health Panel, Issue Brief 2, March 2007. As of September 12, 2016:
http://www.rupri.org/Forms/RuralDefinitionsBrief.pdf

Courts Task Force, *Court Systems Information Technology Priorities in the Aftermath of the Events of September 11, 2001*, November 27, 2001. As of August 22, 2016:
http://www.search.org/files/pdf/TF_Recs.pdf

Crank, John P., "Civilianization in Small and Medium Police Departments in Illinois, 1973–1986," *Journal of Criminal Justice*, Vol. 17, No. 3, 1989, pp. 167–177.

DeFrances, Carol J., *State Court Prosecutors in Small Districts, 2001*, Washington, D.C.: Bureau of Justice Statistics, NCJ 196020, January 1, 2003. As of August 22, 2016:
http://www.bjs.gov/index.cfm?ty=pbdetail&iid=1176

Eastern Kentucky University, *National Assessment of Technology and Training for Small and Rural Law Enforcement Agencies (NATTS): A Descriptive Analysis*, Washington, D.C.: U.S. Department of Justice, National Institute of Justice, Office of Science and Technology, December 2002. As of August 22, 2016:
https://www.ncjrs.gov/app/publications/abstract.aspx?ID=198619

Economic Research Service, U.S. Department of Agriculture, "2013 Rural–Urban Continuum Codes," data set, last updated May 10, 2013. As of September 12, 2016:
http://www.ers.usda.gov/data-products/rural-urban-continuum-codes/.aspx

Eisenstein, James, Roy B. Flemming, and Peter F. Nardulli, *The Contours of Justice: Communities and Their Courts*, Boston, Mass.: Little, Brown, 1988.

Ellsworth, Thomas, and Ralph A. Weisheit, "The Supervision and Treatment of Offenders on Probation: Understanding Rural and Urban Differences," *Prison Journal*, Vol. 77, No. 2, June 1997, pp. 209–228.

Falcone, David N., L. Edward Wells, and Ralph A. Weisheit, "The Small-Town Police Department," *Policing*, Vol. 25, No. 2, 2002, pp. 371–384.

Feld, Barry C., "Justice by Geography: Urban, Suburban, and Rural Variations in Juvenile Justice Administration," *Journal of Criminal Law and Criminology*, Vol. 82, No. 1, Spring 1991, pp. 156–210.

Foreman, Kelly, "Patrolling Rural Kentucky: Finding the Tactical Edge in the Commonwealth's Countryside," *Kentucky Law Enforcement*, Vol. 12, No. 3, Fall 2013, pp. 40–51. As of August 22, 2016:
https://docjt.ky.gov/Magazines/Issue%2047/files/assets/common/downloads/untitled.pdf

Frey, Heather E., "Tribal Court CASA: A Guide to Program Development," *OJJDP Fact Sheet*, No. 9, June 2002. As of August 22, 2016:
https://www.ncjrs.gov/pdffiles1/ojjdp/fs200209.pdf

Gallas, Geoff, *Court Technology Survey Report*, National Task Force on Court Automation and Integration, November 2001. As of August 22, 2016:
http://www.search.org/files/pdf/CourtTechnologySurveyRepor.pdf

Gardner, Jerry, "Improving the Relationship Between Indian Nations, the Federal Government, and State Governments," *Tribal Court Clearinghouse: A Project of the Tribal Law and Policy Institute*, undated. As of August 26, 2016:
http://www.tribal-institute.org/articles/mou.htm

Garicano, Luis, and Paul Heaton, "Information Technology, Organization, and Productivity in the Public Sector: Evidence from Police Departments," *Journal of Labor Economics*, Vol. 28, No. 1, January 2010, pp. 167–201.

Garwin, Thomas M., Neal A. Pollard, and Robert V. Tuohy, eds., *Project Responder: National Technology Plan for Emergency Response to Catastrophic Terrorism*, National Memorial Institute for the Prevention of Terrorism and U.S. Department of Homeland Security, April 2004. As of August 22, 2016:
https://www.fsi.illinois.edu/documents/research/project-responder-national-technology-plan.pdf

Goldberg, Carole, and Duane Champagne, *Final Report: Law Enforcement and Criminal Justice Under Public Law 280*, Washington, D.C.: U.S. Department of Justice, National Institute of Justice, Office of Justice Programs, November 1, 2007. As of August 22, 2016:
https://www.ncjrs.gov/pdffiles1/nij/grants/222585.pdf

Goodison, Sean E., Robert C. Davis, and Brian A. Jackson, *Digital Evidence and the U.S. Criminal Justice System: Identifying Technology and Other Needs to More Effectively Acquire and Utilize Digital Evidence*, Santa Monica, Calif.: RAND Corporation, RR-890-NIJ, 2015. As of August 22, 2016:
http://www.rand.org/pubs/research_reports/RR890.html

Gordon, John, IV, Brett Andrew Wallace, Daniel Tremblay, and John S. Hollywood, *Keeping Law Enforcement Connected: Information Technology Needs from State and Local Agencies*, Santa Monica, Calif.: RAND Corporation, TR-1165-NIJ, 2012. As of August 22, 2016:
http://www.rand.org/pubs/technical_reports/TR1165.html

Hill, James L., *Investigating Factors That Influence Trial Court Judges' Adoption of Technology*, Case Western Reserve University, doctoral thesis, July 2004. As of August 22, 2016:
https://digital.case.edu/concern/texts/ksl:weaedm091

Hollywood, John S., John E. Boon Jr., Richard Silberglitt, Brian G. Chow, and Brian A. Jackson, *High-Priority Information Technology Needs for Law Enforcement*, Santa Monica, Calif.: RAND Corporation, RR-737-NIJ, 2015. As of August 22, 2016:
http://www.rand.org/pubs/research_reports/RR737.html

Hollywood, John S., and Zev Winkelman, *Improving Information-Sharing Across Law Enforcement: Why Can't We Know?* Santa Monica, Calif.: RAND Corporation, RR-645-NIJ, 2015. As of August 22, 2016:
http://www.rand.org/pubs/research_reports/RR645.html

Hollywood, John S., Dulani Woods, Richard Silberglitt, and Brian A. Jackson, *Using Future Internet Technologies to Strengthen Criminal Justice*, Santa Monica, Calif.: RAND Corporation, RR-928-NIJ, 2015. As of August 22, 2016:
http://www.rand.org/pubs/research_reports/RR928.html

Homeland Security Studies and Analysis Institute, *Project Responder 3: Toward the First Responder of the Future*, Washington, D.C.: U.S. Department of Homeland Security, Science and Technology Directorate, March 2012. As of August 22, 2016:
http://www.nisconsortium.org/portal/resources/bin/Project_Responder_3:_1423591018.pdf

IACP—*See* International Association of Chiefs of Police.

International Association of Chiefs of Police, *Law Enforcement Priorities for Public Safety: Identifying Critical Technology Needs—Technology Survey Results*, Alexandria, Va.: International Association of Chiefs of Police, Fall 2005. As of August 22, 2016:
http://www.theiacp.org/portals/0/pdfs/TechSurveyReport.pdf

———, *Police Chiefs Guide to Immigration Issues*, July 2007. As of August 22, 2016:
http://www.theiacp.org/Portals/0/pdfs/Publications/PoliceChiefsGuidetoImmigration.pdf

———, *Identifying Critical Technology Needs: Technology Survey Results Fall 2005*, November 3, 2008.

Jackson, Brian A., Joe Russo, John S. Hollywood, Dulani Woods, Richard Silberglitt, George B. Drake, John S. Shaffer, Mikhail Zaydman, and Brian G. Chow, *Fostering Innovation in Community and Institutional Corrections: Identifying High-Priority Technology and Other Needs for the U.S. Corrections Sector*, Santa Monica, Calif.: RAND Corporation, RR-820-NIJ, 2015. As of August 22, 2016:
http://www.rand.org/pubs/research_reports/RR820.html

Jones, B. J., *Role of Indian Tribal Courts in the Justice System*, March 2000. As of August 22, 2016:
http://www.icctc.org/Tribal%20Courts.pdf

Justice Management Institute, "Rural Court Information Network," undated. As of August 26, 2016:
http://www.jmijustice.org/network-coordination/rural-court-information-network/

Justice Technology Information Center, "Justice Innovation Center for Small, Rural, Tribal, and Border Criminal Justice Agencies," undated. As of October 31, 2016:
https://www.justnet.org/about/jic-center.html

Kellar, M., *Texas County Jails, 2001: A Status Report*, Texas Commission on Jail Standards, 2001. As of August 22, 2016:
http://www.tcjs.state.tx.us/docs/Final DraftTJSbu.pdf

King, Ryan S., Marc Mauer, and Tracy Huling, *Big Prisons, Small Towns: Prison Economics in Rural America*, Washington, D.C.: Sentencing Project, February 2003. As of August 22, 2016:
http://www.sentencingproject.org/wp-content/uploads/2016/01/Big-Prisons-Small-Towns-Prison-Economics-in-Rural-America.pdf

Koper, Christopher S., Cynthia Lum, and James J. Willis, "Optimizing the Use of Technology in Policing: Results and Implications from a Multi-Site Study of the Social, Organizational, and Behavioural Aspects of Implementing Police Technologies," *Policing*, Vol. 8, No. 2, 2014, pp. 212–221.

Koper, Christopher S., Bruce G. Taylor, and Bruce E. Kubu, *Law Enforcement Technology Needs Assessment: Future Technologies to Address the Operational Needs of Law Enforcement*, Washington, D.C.: Police Executive Research Forum, January 16, 2009. As of August 22, 2016:
http://www.policeforum.org/assets/docs/Free_Online_Documents/Technology/law%20enforcement%20technology%20needs%20assessment%202009.pdf

Kuhns, Joseph B., III, Edward R. Maguire, and Stephen M. Cox, "Public-Safety Concerns Among Law Enforcement Agencies in Suburban and Rural America," *Police Quarterly*, Vol. 10, No. 4, December 2007, pp. 429–454.

Lambert, John T., "Attorneys and Their Use of Technology," *Entrepreneurial Executive*, Vol. 13, 2008, pp. 83–99.

LaTourrette, Tom, D. J. Peterson, James T. Bartis, Brian A. Jackson, and Ari Houser, *Protecting Emergency Responders*, Vol. 2: *Community Views of Safety and Health Risks and Personal Protection Needs*, Santa Monica, Calif.: RAND Corporation, MR-1646-NIOSH, 2003. As of August 22, 2016:
http://www.rand.org/pubs/monograph_reports/MR1646.html

Law Enforcement and Corrections Technology Advisory Council, *2009 Law Enforcement and Corrections Technology Advisory Council Annual Report*, September 2009. As of August 22, 2016:
https://www.justnet.org/pdf/2009-LECTAC-Report.pdf

Leadership Directories, *Directory of State Court Clerks and County Courthouses: 2014 Edition*, September 2013.

LECTAC—*See* Law Enforcement and Corrections Technology Advisory Council.

Liederbach, John, and James Frank, "Policing Mayberry: The Work Routines of Small-Town and Rural Officers," *American Journal of Criminal Justice*, Vol. 28, No. 1, September 2003, pp. 53–72.

LIS, *Technology Issues in Corrections Agencies: Results of a 1995 Survey*, Longmont, Colo.: U.S. Department of Justice, National Institute of Corrections Information Center, July 1995. As of August 22, 2016:
http://static.nicic.gov/Library/012377.pdf

Logan, T. K., Lisa Shannon, and Robert Walker, "Protective Orders in Rural and Urban Areas: A Multiple Perspective Study," *Violence Against Women*, Vol. 11, No. 7, July 2005, pp. 876–911.

Maguire, Edward R., Joseph B. Kuhns, Craig D. Uchida, and Stephen M. Cox, "Patterns of Policing in Nonurban America," *Journal of Research in Crime and Delinquency*, Vol. 34, No. 3, August 1997, pp. 368–394.

Mahoney, Barry, Alan Carlson, and Aimee Baehler, "Strengthening Rural Courts: What Should Be Done to Improve Court Operations and Enhance the Quality of Justice in Rural America?" *Court Manager*, Vol. 20, No. 4, Winter 2005–2006, pp. 11–16.

Maupin, James R., and Lisa J. Bond-Maupin, "Detention Decision-Making in a Predominantly Hispanic Region: Rural and Non-Rural Differences," *Juvenile and Family Court Journal*, Vol. 50, No. 3, July 1999, pp. 11–23.

Mays, G. Larry, and Joel A. Thompson, "Mayberry Revisited: The Characteristics and Operations of America's Small Jails," *Justice Quarterly*, Vol. 5, No. 3, 1988, pp. 421–440.

Melton, Ada, Kimberly Cobb, Adrienne Lindsey, R. Brian Colgan, and David J. Melton, "Addressing Responsivity Issues with Criminal Justice–Involved Native Americans," *Federal Probation*, Vol. 78, No. 2, September 2014, pp. 24–31. As of August 22, 2016:
http://www.uscourts.gov/file/3305/download

Melton, Ada Pecos, Roshanna Lucero, and David J. Melton, *Strategies for Creating Offender Reentry Programs in Indian Country*, Washington, D.C.: U.S. Department of Justice, Office of Justice Programs, August 2010. As of August 22, 2016:
http://www.aidainc.net/Publications/Full_Prisoner_Reentry.pdf

Metoui, Jessica, "Returning to the Circle: The Reemergence of Traditional Dispute Resolution in Native American Communities," *Journal of Dispute Resolution*, Vol. 2007, No. 2, 2007, pp. 517–539. As of August 22, 2016:
http://scholarship.law.missouri.edu/jdr/vol2007/iss2/6/

Minton, Todd D., *Jails in Indian Country, 2014*, Washington, D.C.: Bureau of Justice Statistics, NCJ 248974, October 25, 2015. As of August 22, 2016:
http://www.bjs.gov/index.cfm?ty=pbdetail&iid=5414

Minton, Todd D., Scott Ginder, Susan M. Brumbaugh, Hope Smiley McDonald, and Harley Rohloff, *Census of Jails: Population Changes, 1999–2013*, Washington, D.C.: Bureau of Justice Statistics, NCJ 248627, December 8, 2015. As of August 22, 2016:
http://www.bjs.gov/index.cfm?ty=pbdetail&iid=5480

Minton, Todd D., and Zhen Zeng, *Jail Inmates at Midyear 2014*, Washington, D.C.: Bureau of Justice Statistics, NCJ 248629, June 11, 2015. As of August 22, 2016:
http://www.bjs.gov/index.cfm?ty=pbdetail&iid=5299

Mirsky, Laura, *Restorative Justice Practices of Native American, First Nation and Other Indigenous People of North America*, Part One, April 27, 2004a. As of August 22, 2016:
http://www.iirp.edu/article_detail.php?article_id=NDA1

———, *Restorative Justice Practices of Native American, First Nation and Other Indigenous People of North America*, Part Two, May 26, 2004b. As of August 22, 2016:
http://www.iirp.edu/article_detail.php?article_id=NDA0

NAICJA—*See* National American Indian Court Judges Association.

National American Indian Court Judges Association, "National Tribal Justice Resource Center," undated.

National Center for State Courts, "Rural Courts: Resource Guide," undated (a). As of August 26, 2016:
http://www.ncsc.org/Topics/Special-Jurisdiction/Rural-Courts/Resource-Guide.aspx

———, "Technology in the Courts: Resource Guide," undated (b). As of August 26, 2016:
http://www.ncsc.org/Topics/Technology/Technology-in-the-Courts/Resource-Guide.aspx

———, "Tribal Courts: Resource Guide," undated (c). As of August 26, 2016:
http://www.ncsc.org/Topics/Special-Jurisdiction/Tribal-Courts/Resource-Guide.aspx

———, "Review of NCSC Tribal Projects," December 2009. As of August 22, 2016:
https://www.ncsc.org/~/media/Files/PDF/Jury/Tribal%20Courts%20Project%20Review.ashx

———, *Reengineering Rural Justice in Minnesota's Eighth Judicial District: A Case Study—Improving Efficiencies, Reducing Costs, and Enhancing Operations in Rural Courts: Final Report*, October 2010. As of August 22, 2016:
http://www.ncsc.org/~/media/Files/PDF/Services%20and%20Experts/Court%20reengineering/Minnesota%20Reengineering%20Final%20Report.ashx

National Institute of Justice, "Law Enforcement Technology: Are Small and Rural Agencies Equipped and Trained?" *Research for Practice*, June 2004. As of August 22, 2016:
https://www.ncjrs.gov/pdffiles1/nij/204609.pdf

———, *High-Priority Criminal Justice Technology Needs*, NCJ 225375, March 2009. As of August 22, 2016:
https://www.ncjrs.gov/pdffiles1/nij/225375.pdf

National Law Enforcement and Corrections Technology Center, *United States–Canadian Border Summit*, 2011. As of August 22, 2016:
https://www.justnet.org/pdf/00-Border-Cover-Lowres.pdf

Native American Rights Fund, "National Indian Law Library," undated. As of August 26, 2016:
http://www.narf.org/nill/

NCSC—*See* National Center for State Courts.

NIJ—*See* National Institute of Justice.

NLECTC—*See* National Law Enforcement and Corrections Technology Center.

Nugent-Borakove, Elaine, Barry Mahoney, and Debra Whitcomb, "Strengthening Rural Courts: Challenges and Progress," *Future Trends in State Courts*, National Center for State Courts, 2011. As of August 22, 2016:
http://www.ncsc.org/sitecore/content/microsites/future-trends-2011/home/Specialized-Courts-Services/3-2-Strengthening-Rural-Courts.aspx

Nunn, Samuel, "How Capital Technologies Affect Municipal Service Outcomes: The Case of Police Mobile Digital Terminals and Stolen Vehicle Recoveries," *Journal of Policy Analysis and Management*, Vol. 13, No. 3, Summer 1994, pp. 539–559.

———, "Police Information Technology: Assessing the Effects of Computerization on Urban Police Functions," *Public Administration Review*, Vol. 61, No. 2, March–April 2001, pp. 221–234.

OJP—*See* Office of Justice Programs.

Olson, David E., Ralph A. Weisheit, and Thomas Ellsworth, "Getting Down to Business: A Comparison of Rural and Urban Probationers, Probation Sentences, and Probation Outcomes," *Journal of Contemporary Criminal Justice*, Vol. 17, No. 1, February 2001, pp. 4–18.

Owens, Colleen, Meredith Dank, Justin Breaux, Isela Bañuelos, Amy Farrell, Rebecca Pfeffer, Katie Bright, Ryan Heitsmith, and Jack McDevitt, *Understanding the Organization, Operation, and Victimization Process of Labor Trafficking in the United States*, Washington, D.C.: Urban Institute, October 2014. As of August 22, 2016:
http://www.urban.org/sites/default/files/alfresco/publication-pdfs/413249-Understanding-the-Organization-Operation-and-Victimization-Process-of-Labor-Trafficking-in-the-United-States.PDF

Perin, Michelle, "First Responder, Last Frontier: Alaska's Village Public Safety Officers," *Law Enforcement Technology*, Vol. 41, August 25, 2014. As of August 22, 2016:
http://www.officer.com/article/11535791/law-enforcement-in-alaska

Perry, Steven W., *American Indians and Crime*, Washington, D.C.: U.S. Department of Justice, Office of Justice Programs, Bureau of Justice Statistics, NCJ 203097, December 2004. As of August 22, 2016:
http://www.bjs.gov/content/pub/pdf/aic02.pdf

———, *Census of Tribal Justice Agencies in Indian Country, 2002*, Washington, D.C.: Bureau of Justice Statistics, NCJ 205332, December 1, 2005. As of August 22, 2016:
http://www.bjs.gov/index.cfm?ty=pbdetail&iid=543

Perry, Steven W., and Duren Banks, "Prosecutors in State Courts, 2007: Statistical Tables," *2007 National Census of State Court Prosecutors*, NCJ 234211, December 2011. As of August 25, 2016:
http://www.bjs.gov/content/pub/pdf/psc07st.pdf

Police Executive Research Forum, "Use of Technology in Policing: The Chief's Perspective," briefing, Washington D.C., April 4, 2011.

———, *How Are Innovations in Technology Transforming Policing?* Washington, D.C., January 2012.

Pruitt, Lisa R., J. Cliff McKinney II, Juliana Fehrenbacher, and Amy Dunn Johnson, "Access to Justice in Rural Arkansas," Arkansas Access to Justice Commission, March 2015. As of August 22, 2016:
http://www.arkansasjustice.org/sites/default/files/file%20attachments/AATJPolicyBrief2015-0420.pdf

Public Law 83-280, conferring criminal and civil jurisdiction on California, Minnesota, Nebraska, Oregon, and Wisconsin for causes of action arising on reservations in those states, August 15, 1953. As of August 24, 2016:
https://www.gpo.gov/fdsys/pkg/STATUTE-67/pdf/STATUTE-67-Pg588.pdf

Public Law 93-638, Indian Self-Determination and Education Assistance Act of 1975, January 4, 1975.

Public Law 95-608, Indian Child Welfare Act of 1978, November 8, 1978. As of August 24, 2016:
https://www.gpo.gov/fdsys/pkg/STATUTE-92/pdf/STATUTE-92-Pg3069.pdf

Public Law 108-79, Prison Rape Elimination Act of 2003, September 4, 2003. As of August 29, 2016:
https://www.gpo.gov/fdsys/pkg/PLAW-108publ79/pdf/PLAW-108publ79.pdf

Public Law 111-211, Tribal Law and Order Act, July 29, 2010. As of August 24, 2016:
https://www.gpo.gov/fdsys/pkg/PLAW-111publ211/html/PLAW-111publ211.htm

Race, Melanie M., Anush Yousefian, David Lambert, and David Hartley, *Mental Health Services in Rural Jails*, Maine Rural Health Research Center, Working Paper 42, August 2010. As of July 8, 2016:
http://muskie.usm.maine.edu/Publications/rural/Rural-Jails-Mental-Health.pdf

Reaves, Brian A., *Census of State and Local Law Enforcement Agencies, 2008*, Washington, D.C.: U.S. Department of Justice, Bureau of Justice Statistics, NCJ 233982, July 26, 2011. As of August 24, 2016: http://www.bjs.gov/index.cfm?ty=pbdetail&iid=2216

———, *Local Police Departments, 2013: Personnel, Policies, and Practices*, Washington, D.C.: U.S. Department of Justice, Bureau of Justice Statistics, NCJ 248677, May 2015. As of August 22, 2016: http://www.bjs.gov/content/pub/pdf/lpd13ppp.pdf

Rodriguez, Nancy, *A Multilevel Analysis of Juvenile Court Processes: The Importance of Community Characteristics*, National Institute of Justice, June 30, 2008. As of August 22, 2016: https://www.ncjrs.gov/pdffiles1/nij/grants/223465.pdf

Romesburg, William H., *Law Enforcement Tech Guide for Small and Rural Police Agencies: A Guide for Executives, Managers, and Technologists*, Washington, D.C.: U.S. Department of Justice, Office of Community Oriented Policing Services, NCJ 211995, 2005. As of August 22, 2016: http://www.search.org/files/pdf/SmallRuralTechGuide.pdf

Royal, Michelle, Amy Donahue, and Aidan Kirby, *Project Responder: Review of Emergency Response Capability Needs*, National Memorial Institute for the Prevention of Terrorism and U.S. Department of Homeland Security Report, February 2008.

Ruddell, Rick, and G. Larry Mays, "Expand or Expire: Jails in Rural America," *Corrections Compendium*, Vol. 31, No. 6, November–December 2006, pp. 1–4.

Sale, Jeffry, *National Summit on Small and Rural Law Enforcement 2009*, Hazard, Ky.: Rural Law Enforcement Technology Center, NCJ 233949, August 2009.

———, *Report on the National Small and Rural Agency Summit*, Fort Myers, Fla.: Small, Rural, Tribal and Border Regional Center, August 2010.

Schwabe, William, Lois M. Davis, and Brian A. Jackson, *Challenges and Choices for Crime-Fighting Technology: Federal Support of State and Local Law Enforcement*, Santa Monica, Calif.: RAND Corporation, MR-1349-OSTP/NIJ, 2001. As of August 22, 2016: http://www.rand.org/pubs/monograph_reports/MR1349.html

Scism, Bill, "Correctional Security Technology: Catch the Wave," *Corrections Today*, Vol. 71, No. 4, August 2009, pp. 8–12.

Silberglitt, Richard, Brian G. Chow, John S. Hollywood, Dulani Woods, Mikhail Zaydman, and Brian A. Jackson, *Visions of Law Enforcement Technology in the Period 2024–2034: Report of the Law Enforcement Futuring Workshop*, Santa Monica, Calif.: RAND Corporation, RR-908-NIJ, 2015. As of August 22, 2016: http://www.rand.org/pubs/research_reports/RR908.html

Sims, Victor H., *Small Town and Rural Police*, Springfield, Ill.: Charles C. Thomas, 1988.

Skogan, Wesley G., and Susan M. Hartnett, "The Diffusion of Information Technology in Policing," *Police Practice and Research*, Vol. 6, No. 5, 2005, pp. 401–417.

SocioCultural Research Consultants, *Dedoose*, version 6.1.18, 2015.

Spangenberg, Robert L., Marea L. Beeman, David J. Carroll, David Freedman, Evelyn Pan, David J. Newhouse, and Dorothy Chan, *Indigent Defense and Technology: A Progress Report*, Washington, D.C.: Bureau of Justice Assistance, NCJ 179003, November 1999. As of August 22, 2016: https://www.ncjrs.gov/pdffiles1/bja/179003.pdf

Stephan, James J., *Census of Jails, 1999*, Washington, D.C.: U.S. Department of Justice, Office of Justice Programs, Bureau of Justice Statistics, NCJ 18633, August 2001. As of August 24, 2016: http://www.bjs.gov/content/pub/pdf/cj99.pdf

Stephan, James, and Georgette Walsh, *Census of Jail Facilities, 2006*, Washington, D.C.: U.S. Department of Justice, Office of Justice Programs, Bureau of Justice Statistics, NCJ 230188, December 2011. As of August 22, 2016: http://www.bjs.gov/content/pub/pdf/cjf06.pdf

Thomson, Doug, and David Fogel, *Probation Work in Small Agencies: A National Study of Training Provisions and Needs*, Vol. 1, Chicago, Ill.: University of Illinois, Chicago, 1980, revised March 1981. As of August 22, 2016:
https://www.ncjrs.gov/pdffiles1/Digitization/90451-90452NCJRS.pdf

Tribal Law and Policy Institute, *Tribal Healing to Wellness Courts: The Key Components*, Washington, D.C.: U.S. Department of Justice, Office of Justice Programs, NCJ 188154, April 2003. As of August 22, 2016:
https://www.ncjrs.gov/pdffiles1/bja/188154.pdf

Ulmer, Jeffery T., and Mindy S. Bradley, "Variation in Trial Penalties Among Serious Violent Offenses," *Criminology*, Vol. 44, No. 3, 2006, pp. 631–670.

Uniform Crime Reporting Program, "2014 Crime in the United States," undated. As of September 12, 2016:
https://ucr.fbi.gov/crime-in-the-u.s/2014/crime-in-the-u.s.-2014/cius-home

U.S. Department of Interior, Office of Inspector General, *"Neither Safe nor Secure": An Assessment of Indian Detention Facilities*, Report 2004-I-0056, September 2004. As of August 22, 2016:
https://www.doioig.gov/reports/neither-safe-nor-secure-assessment-indian-detention-facilities

———, *Bureau of Indian Affairs' Detention Facilities*, Report WR-EV-BIA-0005-2010, March 2011. As of August 22, 2016:
https://www.doioig.gov/sites/doioig.gov/files/01-WR-EV-BIA-0005-2010Public.pdf

Vetter, Stephanie J., and John Clark, *The Delivery of Pretrial Justice in Rural Areas: A Guide for Rural County Officials*, Washington, D.C.: Pretrial Justice Institute and National Association of Counties, 2013. As of August 22, 2016:
http://www.naco.org/sites/default/files/documents/The%20Delivery%20of%20Pretrial%20Justice%20in%20Rural%20Areas%20-%20A%20Guide%20for%20Rural%20County%20Officials.pdf

Wakeling, Stewart, Miriam Jorgensen, Susan Michaelson, and Manley Begay, *Policing on American Indian Reservations: A Report to the National Institute of Justice*, Washington, D.C.: U.S. Department of Justice, Office of Justice Programs, National Institute of Justice, NCJ 188095, July 2001. As of August 23, 2016:
https://www.ncjrs.gov/pdffiles1/nij/188095.pdf

Weisburd, David, and Cynthia Lum, "The Diffusion of Computerized Crime Mapping in Policing: Linking Research and Practice," *Police Practice and Research*, Vol. 6, No. 5, 2005, pp. 419–434.

Weisheit, Ralph A., David N. Falcone, and L. Edward Wells, *Rural Crime and Rural Policing*, Washington, D.C.: U.S. Department of Justice, Office of Justice Programs, National Institute of Justice, NCJ 150223, September 1994. As of August 23, 2016:
https://www.ncjrs.gov/pdffiles/rcrp.pdf

Wells, L. Edward, and David N. Falcone, "Tribal Policing on American Indian Reservations," *Policing*, Vol. 31, No. 4, 2008, pp. 648–673.

Welsh, Brendan C., "Technological Innovations for Policing: Crime Prevention as the Bottom Line," *Criminology and Public Policy*, Vol. 2, No. 1, November 2002, pp. 129–132.

Wilcox, Pamela, Carol E. Jordan, Adam J. Pritchard, and Ryan Randa, "Rurality–Urbanism and Protective Order Service: A Research Note," *Journal of Crime and Justice*, Vol. 31, No. 2, 2008, pp. 65–86.

Wodahl, Eric J., "The Challenges of Prisoner Reentry from a Rural Perspective," *Western Criminology Review*, Vol. 7, No. 2, 2006, pp. 32–47.

Wyoming Center for Legal Aid, "Access to Justice Commission Receives Grant from the ABA," *Legal Hub*, Vol. 1, No. 3, July 25, 2013. As of August 23, 2016:
http://www.legalhelpwy.org/index.php/about-us/news/489/

Zelenak, Anthony, and H. Buford Goff Jr., "Security and Technology: The Past 30 Years," *Corrections Today*, Vol. 67, No. 4, July 2005, pp. 52–55, 98.